HIGHER PHYSICS

Hugh McGill

Hodder Gibson

A MEMBER OF THE HODDER HEADLINE GROUP

Every effort has been made to trace all copyright holders, but if any have been inadvertently overlooked the Publishers will be pleased to make the necessary arrangements at the first opportunity.

Although every effort has been made to ensure that website addresses are correct at time of going to press, Hodder Gibson cannot be held responsible for the content of any website mentioned in this book. It is sometimes possible to find a relocated web page by typing in the address of the home page for a website in the URL window of your browser.

Papers used in this book are natural, renewable and recyclable products. They are made from wood grown in sustainable forests. The logging and manufacturing processes conform to the environmental regulations of the country of origin.

Orders: please contact Bookpoint Ltd, 130 Milton Park, Abingdon, Oxon OX14 4SB. Telephone: (44) 01235 827720. Fax: (44) 01235 400454. Lines are open from 9.00 - 6.00, Monday to Saturday, with a 24-hour message answering service. Visit our website at www.hoddereducation.co.uk. Hodder Gibson can be contacted direct on: Tel: 0141 848 1609; Fax: 0141 889 6315; email: hoddergibson@hodder.co.uk

First published in 2005 by
Hodder Gibson, a member of the Hodder Headline Group
2a Christie Street
Paisley PA1 1NB

Impression number 10 9 8 7 6 5 4 3 2 1
Year 2010 2009 2008 2007 2006 2005

Cover photo by Science Photo Library
Typeset in 9.5 on 12.5pt Frutiger Light by Phoenix Photosetting, Chatham, Kent
Printed and bound in Great Britain by Arrowsmith, Bristol

A catalogue record for this title is available from the British Library

ISBN 10: 0 340 88794X
ISBN 13: 978 0 340 887943

CONTENTS

Contents

WELCOME

Welcome to this revision book!

Well done, you have decided to sit Higher Physics! Good decision. This book is designed to help you do as well as you can.

Higher Physics is enjoyable, interesting and relevant to life in the twenty-first century. It is also a very useful qualification if you want to apply to university or college or if you want to get a job.

The examination is fair. Prepare properly for it and you can expect to do well.

Good luck!!

How to use this book

This book covers everything you need to know for Higher Physics. It also includes tips and hints about how to answer exam questions and about what SQA expects you to be able to do.

The book has five sections.

The first three sections cover the knowledge and understanding of the three Higher Physics units: Mechanics and Properties of Matter, Electricity and Electronics, and Radiation and Matter.

The fourth section covers what you need to know about units and uncertainties.

Each of the first four sections is written in topics. At the start of each topic I have included a summary of what SQA expects you to be able to do and these are labelled What You Should Know. You can use these summaries to help you prepare a study checklist for your prelim or for the final examination.

Study the topics in any order you like but give your brain a chance – study one topic at a time.

Some of the topics include examples that show you how to answer questions. All of the topics include questions for you to try. Just as in the final examination some of the questions are straightforward and some are difficult.

Detailed answers to all of the questions are included in the fifth section of this book. Try to answer the questions without looking up the answers. If you do not know how to answer a question the answers are there to help you. Use the answers wisely.

Make sure you study all of the topics and try all of the questions before you sit the final examination.

The Higher Physics examination

The Higher Physics examination lasts for 2½ hours and is marked out of a total of 90 marks. The marks are spread evenly over the three units of the course.

At the start of the paper there are 20 multiple-choice questions each worth 1 mark. The questions for the remaining 70 marks require written responses. Your answers to these questions could be a few words, a few sentences or numerical calculations.

Approximately 36 marks (40%) are allocated to knowledge and understanding questions.

Knowledge and understanding questions test your ability to use:

- physical quantities and their units
- mathematical relationships to solve straightforward numerical questions
- physics principles to explain familiar observations
- models, e.g. the nuclear model of the atom, in explanations of familiar situations.

In your Higher it is very important that you get as many of the knowledge and understanding questions correct as possible. This will give you a solid basis for passing the exam.

You will only be able to get these marks if you prepare for the examination – so **study properly!**

Approximately 54 marks (60%) are allocated to problem solving questions.

Problem solving questions test your ability to:

- Select and present information.
- Process information to solve numerical questions set in familiar and less familiar situations.
- Draw valid conclusions from information given in familiar and less familiar contexts.
- Explain observations in less familiar situations.
- Plan, design and evaluate experimental procedures.
- Integrate your knowledge and understanding and your problem solving skills across course units.

About half of the problem solving questions (i.e. about 27 marks' worth) will be set in situations that should be reasonably familiar.

The remaining questions may be set in contexts that you have never seen before. They may be also less structured or more complex. These questions are generally the most difficult in the Higher Physics exam. **Do not panic.** You can easily pass your Higher without getting marks for these questions. You do need to get marks in these questions if you want to get an A.

Sitting the examination

Always read the question carefully. Make sure you answer the question that is asked.

Remember, the more you practise the better you get.

Multiple-choice questions

Do not spend too long on these questions – 30 to 35 minutes should leave you enough time to complete the rest of the paper.

Do not expect the multiple-choice questions to be easy. Some will be straightforward but there will also be questions that are difficult.

In Higher Physics there is **one and only one** correct answer to a multiple-choice question.

There are **four wrong answers** to distract you. The best way to avoid being distracted is to cover the answers when you are reading the question.

Work out an answer before looking at the options given in the question paper. You will be able to do this for most questions. When you have decided what you think the answer is then look at the possible answers.

If you are not sure about a question improve your chances by eliminating any possible answers that you know are definitely wrong. If you are completely stuck, guess. You do not lose anything for getting a multiple-choice question wrong.

Numerical questions

When you are solving numerical problems follow the steps below – this will help you get as many marks as possible.

1 Collect data on the left hand side of the page – use the symbols used by SQA and include the symbol for the quantity you want to find (in a small number of questions you will need to obtain data from the Data Sheet at the front of the examination paper).
2 Check the units for each piece of data – make sure that all data have correct SI units – this makes it easier for you to get the unit correct in the final answer.
3 Select the correct relationship from the list in your Physics Data Booklet – look for a relationship that has only symbols included in the list you made in step 1.
4 Write the relationship in the centre of the page level with the first piece of data.
5 Substitute data in the relationship one number at a time. Do not rearrange the relationship before you substitute. Substitute on the line below the relationship making sure the equals signs are in line – this makes it easier for you to check your working.
6 Carry out the calculation – do this at the right hand side of the page or on your calculator.
7 Write down your answer in the form '*symbol* = number' – again make sure the equals signs are in line.
8 If necessary, round to the correct number of significant figures (keep one extra significant figure in any intermediate values you calculate).
9 Add the correct SI unit for the quantity you have just calculated.

The above method is used in the solutions to all numerical questions in this book. There are detailed explanations of the method in the answers to the first three questions in *Exercise 8 Using the equations of motion*.

Explanations and conclusions

In your Higher Physics paper the number of marks available for each question is shown in **bold** on the right hand side of the page.

The more marks there are available, the more you have to write. Normally each relevant piece of physics is worth ½ mark so to get full marks in a 2 mark question you should expect to include four relevant pieces of information.

To get full marks for explanations and conclusions the information you present must be **relevant**, **well-structured** and **complete**. Correct physics that is irrelevant to the question gets zero marks.

In most explanation questions you need to **apply a physics principle or relationship**. For example, questions about moving objects often involve Newton's laws. Electrical questions often involve rules for series and parallel circuits.

Pay attention to verbs used in questions. 'Describe and explain' and 'describe' on its own are different. 'State' is easier to answer than 'justify'. Most of the verbs commonly used by SQA are used in questions included in the exercises in this book. The answers to these questions indicate the level of detail required for full marks.

Use the language of physics. There are many words in physics that have precise meanings – use these words in your written responses. Paraphrasing often introduces inaccuracy and may cost you marks.

Chapter 1

MECHANICS

1.1 Scalars and vectors

What You Should Know

For Higher Physics you need to be able to:

- define and classify vector quantities and scalar quantities
- find the resultant of two or more vectors
- resolve vectors into two perpendicular components.

A **scalar** is a quantity that has magnitude (size) only. A **vector** is a quantity that has both magnitude and direction.

Mass, distance, speed, kinetic energy, potential energy and pressure are all scalars. Weight, displacement, velocity, acceleration, force and momentum are all vectors.

Exercises

Exercise 1 Scalars and vectors

1 Which of the lists (a) to (h) below contain only scalars?

(a) mass, displacement, speed, kinetic energy
(b) pressure, volume, temperature, mass
(c) electric field strength, velocity, acceleration, force
(d) heat, light energy, distance, power
(e) impulse, upthrust, acceleration, weight
(f) time, electric charge, wavelength, absorbed dose
(g) buoyancy force, momentum, velocity, displacement
(h) refractive index, half-life, frequency, potential energy.

2 Which of the lists (a) to (h) above contain only vectors?

Adding vectors

All vectors can be represented by a straight line with an arrow. The length of the line represents the size of the vector. The direction of the arrow on the line represents the direction of the vector.

The sum of two or more vectors is called the **resultant**.

When you add two perpendicular vectors together it is easiest to use Pythagoras' theorem to get the length of the resultant and the tangent to find the angle for the direction.

Questions and Answers

A man walks 120 m due East and then 50 m due North. Find his resultant displacement.

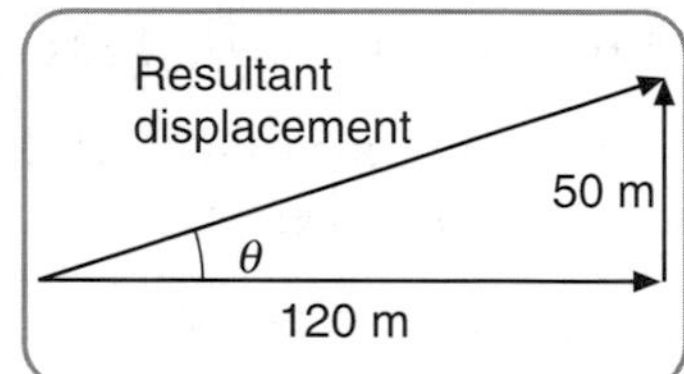

$(\text{resultant})^2 = 120^2 + 50^2 = 16\,900$

resultant $= 130$ m

$\tan\theta = \dfrac{50}{120} = 0{\cdot}417 \Rightarrow \theta = 22{\cdot}6°$

resultant displacement is 130 m 23°N of E

Note: The final answers are given to **two** figures because the minimum of significant figures in the data is two.

When you have to add three or more vectors it may be easier to use a scale diagram. Draw the vectors one after another taking care to get the lengths and angles as accurate as possible.

Vectors are always added 'nose to tail'; that is, the end of the first vector is the starting point for the second vector, and so on. The sum of the vectors is the straight line from the start of the first vector to the end of the last vector. (See Figure 1.1 for example.)

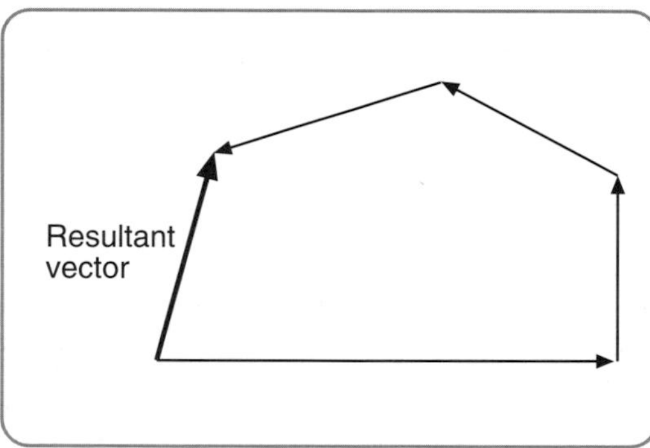

Figure 1.1

Exercises

Exercise 2 Adding vectors

1 A boat sets off at $5{\cdot}0\ \text{m s}^{-1}$ at right angles to the banks of a river. The water in the river is moving at $3{\cdot}5\ \text{m s}^{-1}$. Find the resultant velocity of the boat. (Use Pythagoras and tangent.)

2 A girl runs 300 m West and then 450 m North West. She then runs a further 450 m North East and finally she runs 55 m East. Use a scale drawing to find the girl's resultant displacement.

Components of vectors

Some problems in Higher Physics involve vectors at angles to surfaces. To solve the problem you have to split the vector into two perpendicular vectors, one parallel and one perpendicular to the surface. These vectors are called **components** and they add up to the original vector. Finding the components is called 'resolving' the vector.

In Higher Physics you need to be able to resolve vectors into components in the situations shown below. The text at the side of the diagrams shows you how to work out the components in each situation.

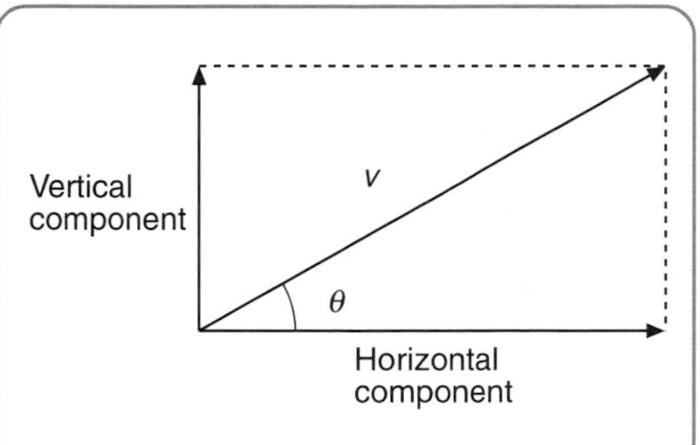

Horizontal and vertical components

To avoid confusion **always** use the angle between the vector and horizontal.

vertical component $= v \sin \theta$

horizontal component $= v \cos \theta$

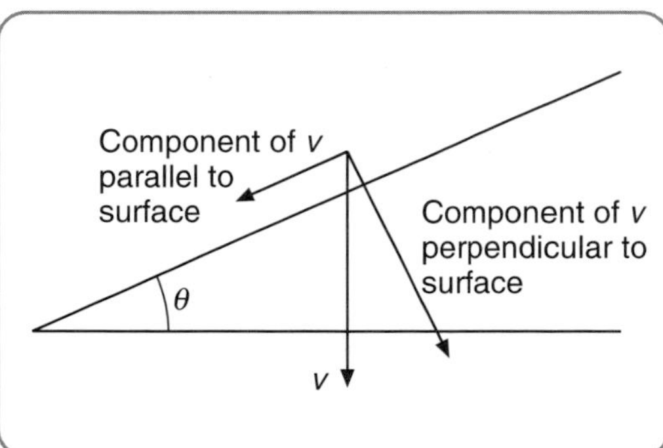

Slopes

To avoid confusion **always** use the angle between the surface and horizontal.

parallel component $= v \sin \theta$

perpendicular component $= v \cos \theta$

Exercises

Exercise 3 Components of vectors

1 A golfer hits a golf ball with an initial velocity of 40·0 m s^{-1} at an angle of 50·0° to the horizontal.

(a) How many significant figures should there be in your answer and why?

(b) Calculate the horizontal component of the initial velocity of the ball.

(c) Calculate the vertical component of the initial velocity of the ball.

2 A trolley of mass 4·6 kg is held stationary on a slope which is at an angle of 37° to the horizontal.

(a) Calculate the component of the weight of the trolley acting down the slope.

(b) Calculate the component of the weight of the trolley acting into the slope.

1.2 Acceleration

What You Should Know

For Higher Physics you need to be able to:

- define acceleration
- describe how to measure acceleration.

When an unbalanced force acts on an object the object **accelerates**. For example an unbalanced force acting on a stationary object makes it move. Similarly an unbalanced force can make a moving object get faster *or* get slower *or* change direction. All of these are accelerations.

Acceleration is defined as the **change in velocity per unit time** $a = \frac{\Delta v}{t}$.

The SI unit of acceleration is the metre per second per second (m s^{-2}).

When an object has constant acceleration from initial velocity *u* to final velocity *v* in a time *t*, the acceleration *a* is given by $a = \frac{v - u}{t}$.

When you use this equation you **must** substitute initial and final velocities in the correct places. Substituting incorrectly is **wrong physics** and you will lose marks.

In Higher Physics, you have to be able to solve problems about objects speeding up or slowing down or both. You do not have to solve numerical problems about objects changing direction.

Exercises

Exercise 4 Acceleration

1 A car starts from rest and reaches a velocity of 25 m s^{-1} after 10 s. Calculate the average acceleration of the car.

2 A cyclist is travelling along a straight horizontal road at 10 m s^{-1}. The cyclist applies the brakes and comes to rest in a time of 4·0 s. Find the acceleration of the cyclist.

3 At the end of a race a runner enters the final straight running at a velocity of 3·4 m s^{-1}. The runner accelerates at 0·20 m s^{-2} for 3·0 s. Calculate the final velocity of the runner.

Experiments

Measuring acceleration

To measure the acceleration of an object you have to measure two velocities, u and v. You also need to measure the time t for the velocity of the object to change from u to v. To measure each velocity you need to measure a displacement and a time.

So to measure the acceleration of an object you need to make **three** time measurements while the object is moving. One of the best ways to do this is to use a computer as shown in Figure 1.2.

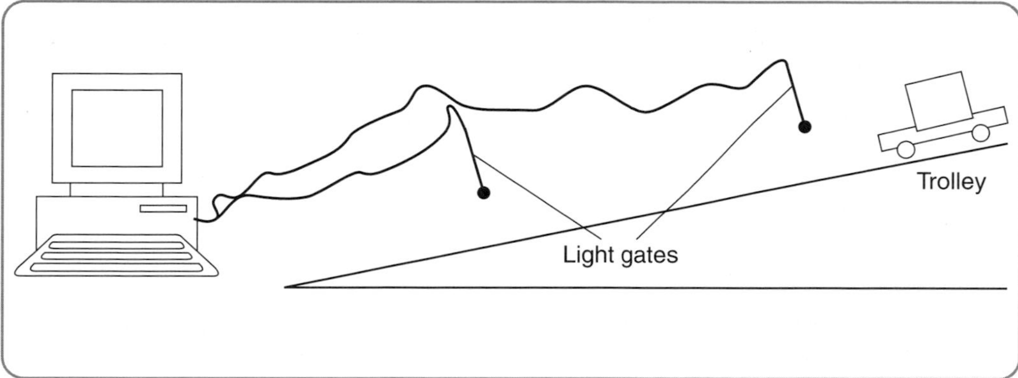

Figure 1.2

Note: A light gate has a light and a sensor that detects when an object is between the light and the sensor.

- Measure the width of a card and attach the card to the trolley.
- Program the computer to measure the time for the card to cut through each of the two light gates and the time for the trolley to travel from the first light gate to the second light gate.
- To find u: divide the width of the card by the time for the card to cut the first light gate.
- To find v: divide the width of the card by the time for the card to cut the second light gate.
- To find a: first work out $(v - u)$; divide this by the time for the trolley to travel from the first light gate to the second light gate.

Exercises

Exercise 5 Measuring acceleration

1 A card of width 0·15 m is attached to a trolley. The trolley is placed at the top of a slope and two light gates are positioned so that the card cuts the gates when the trolley rolls down the slope. The gates are connected to a computer. The trolley is released and the computer records the following time measurements.

time for card to cut first light gate = 0·30 s

time for card to cut second light gate = 0·15 s

time for trolley to travel between light gates = 2·0 s

(a) Calculate the acceleration of the trolley.

(b) How many significant figures should there be in the final answer?

1.3 Graphs of motion

What You Should Know

For Higher Physics you need to be able to:

- draw acceleration–time graphs using information from velocity–time graphs
- use information from graphs and tables to identify constant velocity and constant acceleration.

In Higher Physics three types of graph are used to represent the motion of objects moving in straight lines. These are:

displacement–time (*s*–*t*), **velocity–time (*v*–*t*)** **acceleration–time (*a*–*t*)**.

For constant velocity, the shapes of these graphs are:

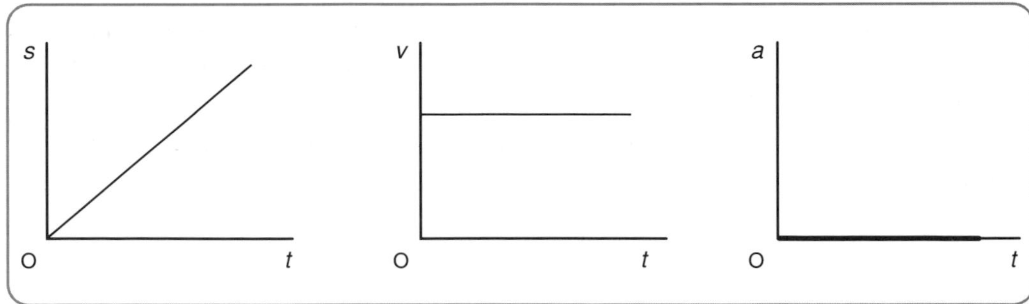

For constant acceleration the shapes of the graphs are:

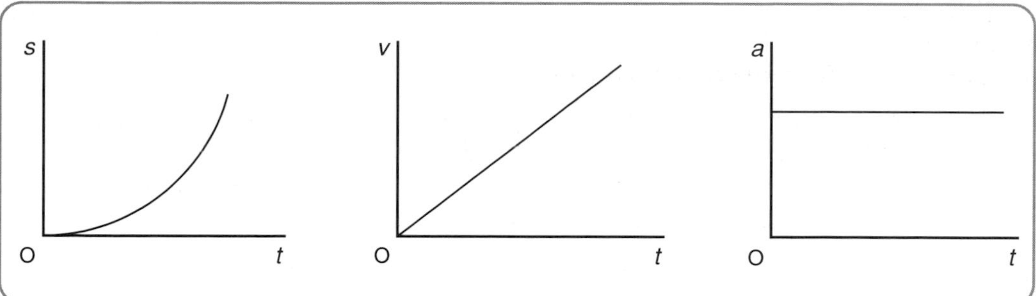

Learn these shapes and the types of motion that they represent.

Notice that the *s–t* graph for constant velocity is the same shape as the *v–t* graph for constant acceleration. Which other two graphs have the same shape?

You are right. The *v–t* graph for constant velocity is the same shape as the *a–t* graph for constant acceleration. Can you think why?

Remember, one way of working out displacement is to calculate the area under a *v–t* graph. A *v–t* graph does not tell you the initial displacement so what you actually work out is the **change** in displacement from the start of the graph. Similarly the area under an *a–t* graph equals the change in velocity from the start of the graph. This is why the shapes of the graphs are similar.

The graphs represent the motion of objects **moving in straight lines**.

When an object moving in a straight line does not change direction its distance and displacement are equal; its *s–t* graph can be labelled either displacement–time or distance–time. Similarly, when an object moving in a straight line does not change direction its speed and velocity are equal, so its *v–t* graph can be labelled either speed–time or velocity–time.

When an object moving in a straight line reverses direction or speeds up and slows down, graphs of its motion have positive and negative sections.

Projectiles

For projectiles you need two graphs – one for horizontal motion and one for vertical motion. When air friction can be ignored, the horizontal motion is always uniform velocity and the vertical motion is always uniform acceleration.

When you are doing projectile questions **always** keep the vertical motion and the horizontal motion separate.

Exercises

Exercise 6 Graphs of motion

1 A car starting from rest accelerates uniformly and reaches a velocity of 24 m s^{-1} after 8·0 s. The car then travels at 24 m s^{-1} for 25 s. The driver then brakes and brings the car to rest in a time of 12 s.

(a) Draw the velocity–time graph of the car's motion. Values are required on both axes.

(b) Using information in the velocity–time graph draw the acceleration–time graph of the car's motion

(c) Calculate the average speed of the car.

2 The velocity–time graph below represents the motion of a bouncing ball released from rest.

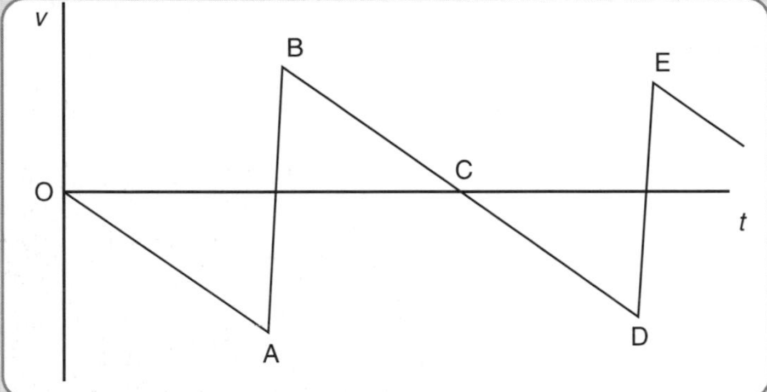

(a) In which parts of the graph is the ball moving downwards?

(b) In which parts of the graph is the ball moving upwards?

(c) Which parts of the graph represent the bounces?

(d) Apart from at O, at which point is the ball stationary in mid-air?

(e) Which parts of the graph represent positive constant acceleration?

(f) Which parts of the graph represent negative constant acceleration?

(g) Sketch the acceleration–time graph for the motion of the ball until point E (values are not required on the axes).

1.4 Equations of motion

What You Should Know

For Higher Physics you need to be able to:

- derive the three equations of motion from basic definitions
- solve problems using the equations of motion.

Deriving the equations of motion

When an object accelerates from initial velocity u to final velocity v in a time t then

the acceleration a is given by $\quad a = \dfrac{v - u}{t}$.

Multiply both sides of this equation by t $\quad \Rightarrow at = v - u$.

Make v the subject of the equation $\quad \Rightarrow v = u + at$. **This is the first equation of motion.**

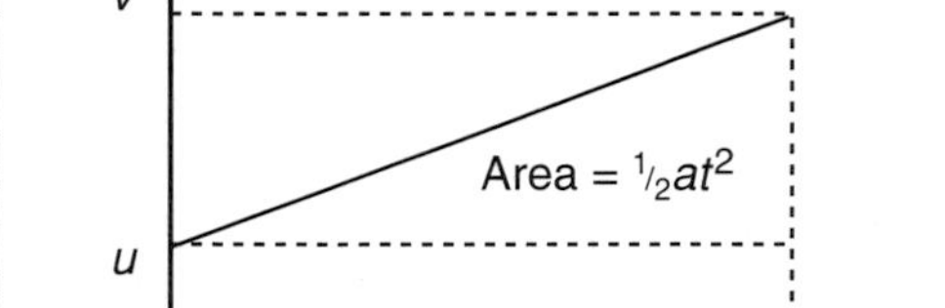

Distance = area under a speed time graph.

area of rectangle $= ut$

area of triangle $= \frac{1}{2} \times (v - u) \times t$

[divide and multiply by t] $= \frac{1}{2} \times \dfrac{(v - u)}{t} \times t^2$

[substitute a] $= \frac{1}{2}at^2$

$\Rightarrow s = ut + \frac{1}{2}at^2$

This is the second equation of motion.

Start with the first equation of motion: $\quad v = u + at$.

Square both sides $\quad \Rightarrow v^2 = u^2 + 2uat + a^2t^2$.

Take $2a$ out of last two terms $\quad \Rightarrow v^2 = u^2 + 2a(ut + \frac{1}{2}at^2)$.

Replace $(ut + \frac{1}{2}at^2)$ with s $\quad \Rightarrow v^2 = u^2 + 2as$.

This is the third equation of motion.

Learn the derivations of these equations.

Exercises

Exercise 7 Deriving the equations of motion

1 Derive the three equations of motion.

(Learn the derivations first – try to write them out without looking at the text above.)

Using the equations of motion

You can use the equations of motion where objects have **uniform acceleration** in a **straight line**. You cannot use the equations of motion where the acceleration is not uniform or where an object is not moving in a straight line. For example, do not use the equations of motion for a problem on a pendulum – the pendulum does not move in a straight line so using the equations of motion would be **wrong physics**.

In Higher Physics, most equation of motion questions can be solved using only one equation. If you find yourself using two equations you are probably not choosing the best equation. This is not wrong and you could still get full marks. However, it does take longer and may waste time during the examination.

Be systematic in the way in which you tackle numerical problems. The solutions to Exercise 8 include explanations of a systematic method. Also, the way the information is set out makes it easier to find errors. Try using this method when you tackle numerical problems.

At first, try the problems without looking at the solutions. If you get stuck, use the solutions at the back of the book – they are there to help you. Remember, the more you practise the better you get.

Exercises

Exercise 8 Using the equations of motion

1 A cyclist moving at 6·4 m s^{-1} pedals harder for 4·0 s and reaches a speed of 7·2 m s^{-1}. Calculate the acceleration of the cyclist.

2 A stone dropped down a deep well hits the water 6·0 s after it is released. How far is the water surface below the top of the well?

3 A boat accelerates uniformly from rest and reaches a velocity of 0·0180 km s^{-1} after travelling 0·250 km. Calculate the acceleration of the boat.

4 (a) Starting with the second equation of motion, show that for an object with uniform acceleration, $s = \frac{1}{2}(u + v)t$.

(b) Hence write an expression for the average velocity of an object with uniform acceleration.

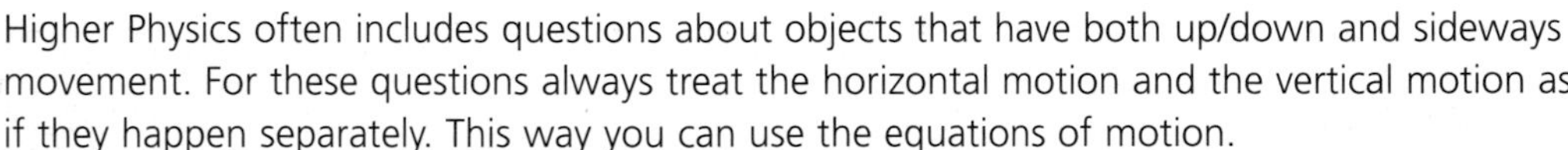

Projectiles

Higher Physics often includes questions about objects that have both up/down and sideways movement. For these questions always treat the horizontal motion and the vertical motion as if they happen separately. This way you can use the equations of motion.

Remember that the time of the horizontal motion is the same as the time for the vertical motion.

For numerical problems on projectiles the horizontal movement is uniform velocity and the vertical motion is uniform acceleration. On Earth the acceleration due to gravity is 9·8 m s^{-2}. Be careful to use the correct value if you get a question about the Moon or a different planet.

Exercises

Exercise 9 Projectiles

1 A golf ball is hit with a velocity of 21 m s^{-1} at an angle of 69° to the horizontal. The ball lands on the fairway in an area level with its starting point. Calculate

(a) the maximum height reached by the golf ball

(b) the time of flight

(c) the horizontal distance travelled by the ball.

1.5 Newton's laws

What You Should Know

For Higher Physics you need to be able to:

- define the newton
- solve problems using $F = ma$
- use diagrams to analyse the forces acting on objects.

Forces and motion

The SI unit of force is the newton. One newton is **defined** as that unbalanced force which causes an acceleration of 1 m s^{-2} when it acts on a mass of 1 kg. Learn this definition.

When the forces acting on an object are **balanced**, the motion of the objects stays the same. **This is Newton's first law**. It means that a stationary object stays stationary. Also, a moving object does not get faster or slower or change direction.

When the forces acting on an object are **unbalanced** the object accelerates – that is, the object gets faster or slower or changes direction.

When you are asked to explain the motion of an object, think about the names, sizes **and** directions of all of the forces. If the object is stationary or moving at a constant speed in a straight line then the forces are balanced. If the motion is changing in any way then the forces are unbalanced.

Exercises

Exercise 10 Balanced and unbalanced forces

1 For each of the following state whether the forces acting are balanced or unbalanced.
  (a) a girl running at a constant speed in a straight line
  (b) a cyclist moving at a constant speed around a corner
  (c) an aeroplane starting its take-off
  (d) a car stopping at traffic lights
  (e) an electron at rest in an electric field
  (f) an electron moving at constant speed in a circle.

Newton's second law

A constant unbalanced force produces a constant acceleration. The direction of the acceleration is the same as the direction of the unbalanced force.

When an unbalanced force acts on an object the acceleration is directly proportional to the unbalanced force and inversely proportional to the mass of the object. **This is Newton's second law**.

Newton's second law is often written as $F = ma$. Remember when you use this equation that F **always** stands for the unbalanced force.

Exercises

Exercise 11 Newton's second law

1 A car of mass 1200 kg has a uniform acceleration of 0·70 m s^{-2}. Calculate the unbalanced force acting on the car.

2 A trolley accelerates uniformly at 500 mm s^{-2} when an unbalanced force of 2·0 N is acting on the trolley. Calculate the mass of the trolley.

3 A cyclist moving at 2·4 m s^{-1} pedals harder for 5·0 s and reaches a speed of 6·4 m s^{-1}. The combined mass of the cyclist and bicycle is 80 kg. Calculate the average unbalanced force acting while the cyclist is accelerating.

Analysing forces

For Higher Physics you have to be able to apply $F = ma$ in situations where more than one force is acting.

In many questions the forces act in only one dimension. For example, in a question about a car pulling a caravan there are three horizontal forces acting on the car: the forward force of the engine, the backward force exerted by the caravan, and backward frictional forces.

Sometimes there are forces in more than one dimension. In this kind of question **always** consider the forces one dimension at a time. Usually in this type of question the forces are balanced in at least one dimension.

A question about an aeroplane in flight could include three vertical forces: the weight of the aeroplane, the upward force due to the wings, and the upward buoyancy force. The question could also include two horizontal forces: the forward force due to the engines and the backward force due to air resistance.

When an aeroplane is flying horizontally the vertical forces are balanced.

Exercises

Exercise 12 Analysing forces

1 A hot air balloon is held stationary by a mooring rope.

(a) Draw a diagram showing the names and directions of all of the vertical forces acting on the balloon.

(b) State the relationship between these forces.

2 A car of mass 1600 kg is towing a caravan of mass 800 kg along a straight horizontal road. At a particular moment the acceleration of the car and caravan is $1{\cdot}50\ \text{m s}^{-2}$.

(a) Calculate the unbalanced force acting on the car.

(b) Calculate the unbalanced force acting on the caravan.

(c) The total frictional forces acting on the car equal 400 N and the total frictional forces acting on the caravan equal 300 N.

(i) Calculate the tension in the coupling between the car and the caravan.

(ii) Calculate the total forward force exerted by the engine.

1.6 Energy and power

What You Should Know

For Higher Physics you need to be able to:

- understand and use the relationships for work done, gravitational potential energy and kinetic energy
- understand and use the relationship $E = Pt$
- solve problems on energy and power.

Work is a process in which energy is converted from one form to another. The SI unit of work and energy is the joule (J). One joule is the work done by a force of 1 N moving through a distance of 1 m.

Power is the rate at which energy is used or work is done. The SI unit of power is the watt (W). One watt is equal to $1\ \text{J s}^{-1}$.

The full name for work done by a force is 'mechanical work done' but this is usually shortened to 'work done'.

Similarly, the term 'gravitational potential energy' is often shortened to 'potential energy'.

Learn the following relationships:

(mechanical) work done:	$E_W = Fd$
(gravitational) potential energy:	$E_p = mgh$
kinetic energy:	$E_k = \frac{1}{2}mv^2$
power:	$P = \frac{E}{t}$ and $E = Pt$

Questions and Answers

A car is initially stationary. The car engine exerts a constant unbalanced force of 180 N and moves the car through a distance of 600 m. The mass of the car is 1200 kg. Show that the kinetic energy gained by the car is equal to the work done by the unbalanced force.

The work done by the unbalanced force = $Fd = 180 \times 600 = 108\,000$ J.

$F = 180$ N	$F = ma$	*Use Newton's second law to find a.*
$m = 1200$ kg	$\Rightarrow 180 = 1200 \times a$	
$a = ?$	$\Rightarrow a = 0{\cdot}15$ m s^{-2}	
$u = 0$	use $v^2 = u^2 + 2as$	*Now use equations of motion to find v^2.*
$s = 600$ m	$\Rightarrow v^2 = 0 + 2 \times 0{\cdot}15 \times 600$	
$a = 0{\cdot}15$ m s^{-2}	$\Rightarrow v^2 = 180$	
$v = ?$		
	$E_k = \frac{1}{2}mv^2$	*Now calculate E_k.*
	$\Rightarrow E_k = \frac{1}{2} \times 1200 \times 180$	
	$= 108\,000$ J	*This equals the work done by the unbalanced force.*

There are lots of energy and power questions in Higher Physics. This is because energy is important in all three course units. Also, because energy is a scalar, energy relationships can be used where objects do not move in straight lines.

In many numerical questions on energy you need to assume that all of the energy put in is converted to useful energy out. Real life is not like this so you could be asked a question where you need to calculate the amount of energy that is wasted.

Exercises

Exercise 13 Energy and power

1 A sphere of mass 12·0 kg is dropped from a height of 15·2 m onto the ground.
   (a) Calculate the gravitational potential energy of the sphere at the start of its motion.
   (b) Calculate the kinetic energy of the sphere just before it hits the ground.
   (c) Calculate the velocity of the sphere just before it hits the ground.
   (d) After it lands the sphere is stationary. What happens to the kinetic energy of the sphere?

2 A pendulum consists of a length of string and a small brass sphere. The brass sphere is pulled to the side and released. At its lowest point the velocity of the brass sphere is 900 mm s^{-1}. Find the maximum height to which the sphere swings above its lowest point.

3 A crate of mass 84·0 kg is dragged up a slope through a distance of 12·9 m by a constant force of 1·20 kN. The slope makes an angle of 18·0° to the horizontal.

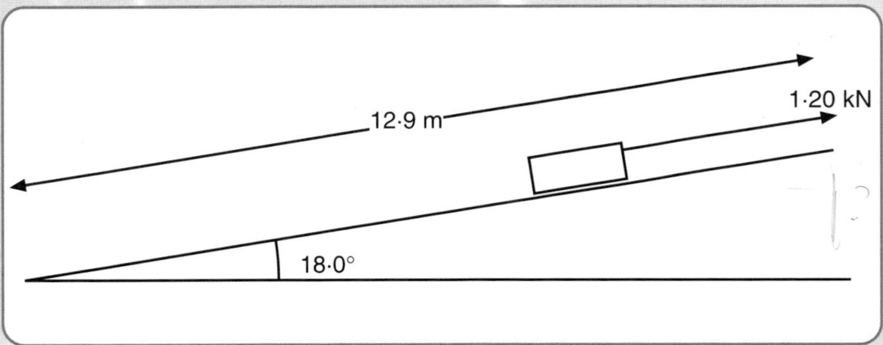

   (a) Calculate the work done by the force.
   (b) Calculate the potential energy gained by the crate.
   (c) Calculate the energy wasted (i.e. the energy converted to forms other than potential energy).
   (d) Calculate the average frictional force between the crate and the slope.
   (e) The time taken to drag the crate up the slope is 80 s. Calculate the
      (i) input power
      (ii) useful power out.

1.7 Momentum and impulse

What You Should Know

For Higher Physics you need to be able to:

- understand and use the quantities momentum and impulse
- state and use the law of conservation of linear momentum
- understand and use the terms elastic collision and inelastic collision
- solve problems on collisions and explosions where objects move in one dimension
- solve problems on impulse.

Momentum

The momentum of an object is the product of its mass and velocity. The SI unit of momentum is kg m s^{-1} and the symbol for momentum used by SQA is p.

$$p = mv$$

Questions and Answers

The momentum of a sphere of mass 6·0 kg is 18 kg m s^{-1}. Calculate the

(a) velocity of the sphere

(b) kinetic energy of the sphere.

(a) $m = 6{\cdot}0$ kg $\quad p = mv$

$p = 18$ kg m s^{-1} $\quad\Rightarrow\quad 18 = 6{\cdot}0 \times v$

$v = ?$ $\quad\Rightarrow\quad v = 3{\cdot}0$ m s^{-1}

(b) $m = 6{\cdot}0$ kg $\quad E_k = \frac{1}{2}mv^2$

$v = 3{\cdot}0$ m s^{-1} $\quad = \frac{1}{2} \times 6{\cdot}0 \times 3{\cdot}0^2$

$= 27$ J

Providing there are no external forces, the **total momentum before** a collision or explosion is equal to the **total momentum after** the collision or explosion. This is the law of conservation of linear momentum.

If you are asked to state this law you must include the condition about external forces to get full marks.

When two objects interact in the absence of external forces, the change in momentum of one is equal and opposite to the change in momentum of the other.

$$(m_1v_1 - m_1u_1) = -(m_2v_2 - m_2u_2)$$

Also, the forces acting on the objects are equal and opposite.

$$F_1 = -F_2$$

You could be asked to prove either or both of the above statements. (The solution to question 4 of *Exercise 14 – Momentum* has the proofs.)

A collision in which total kinetic energy does not change is called an **elastic** collision. A collision in which total kinetic energy changes is called an **inelastic** collision. As explosions result in an increase in kinetic energy explosions are not elastic.

All of the collision and explosion questions in Higher Physics involve objects moving in a single straight line.

Get into the habit of drawing a quick sketch of the situation. Use arrows to show the directions of the velocities and the momenta and write the data given in the question on your diagram.

Questions and Answers

A toy car of mass 2·2 kg with velocity 0·42 m s^{-1} collides with a second toy car of mass 2·2 kg with velocity – 0·20 m s^{-1}. After the collision the two cars stick together.

(a) Calculate the total momentum after the collision.

(b) Hence find the velocity of the cars after the collision.

Before collision | **After collision**

?direction not known?

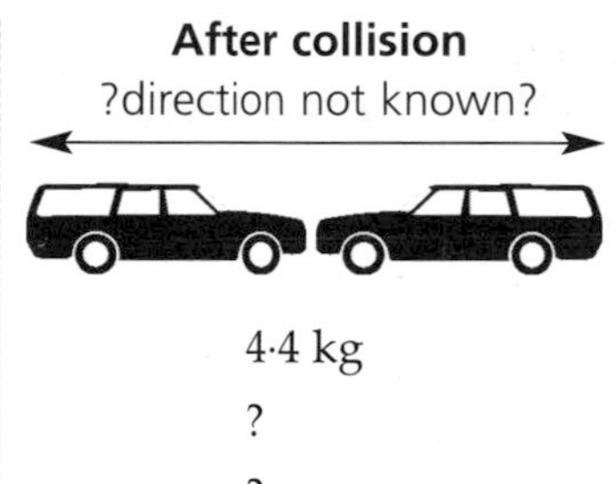

	Before collision		After collision
mass:	2·2 kg	2·2 kg	4·4 kg
velocity:	0·42 m s^{-1}	– 0·20 m s^{-1}	?
momentum:	+ 0·924 kg m s^{-1}	– 0·44 kg m s^{-1}	?

(a) $m_1 = 2{\cdot}2$ kg

$u_1 = 0{\cdot}42$ m s^{-1}

$m_2 = 2{\cdot}2$ kg

$u_2 = -\,0{\cdot}20$ m s^{-1}

$$\text{total momentum before collision} = m_1u_1 + m_2u_2$$
$$= (2{\cdot}2 \times 0{\cdot}42) + (2{\cdot}2 \times -\,0{\cdot}20)$$
$$= +\,0{\cdot}484$$
$$\Rightarrow \text{total momentum after collision} = +\,0{\cdot}48 \text{ kg m s}^{-1}$$

i.e. to the right

(b) mass of cars stuck together = 4·4 kg

$p = 0{\cdot}48$ kg m s^{-1}

$$p = mv$$
$$\Rightarrow \quad 0{\cdot}48 = 4{\cdot}4 \times v$$
$$\Rightarrow \quad v = 0{\cdot}109 = 0{\cdot}11 \text{ m s}^{-1} \text{ to the right}$$

Remember momentum is a **vector** quantity so when you state that a momentum in one direction is positive, any momentum in the opposite direction is negative.

Exercises

Exercise 14 Momentum

1 A male skater of mass 80 kg is skating directly towards a female skater of mass 64 kg. Both skaters are travelling at 2·5 m s^{-1}.

(a) Calculate the total momentum of the two skaters.

(b) State the direction of the total momentum.

2 When the two skaters in question 1 arrive together they both stop. Give a reason why their momentum is not conserved in this 'collision'.

3 A trolley of mass 4·0 kg and with a velocity of 3·0 m s^{-1} collides with a stationary trolley of mass 8·0 kg. Immediately after the collision the 8·0 kg trolley has a velocity of 1·2 m s^{-1}.

(a) Calculate the velocity of the 4·0 kg trolley after the collision.

(b) State whether the collision is elastic or inelastic. You must justify your answer by calculation.

4 An object of mass m_1 and initial velocity u_1 collides with a second object of mass m_2 and initial velocity u_2. After the collision the velocities of the objects are v_1 and v_2 respectively.

(a) Show that the momentum change of m_1 is equal and opposite to the momentum change of m_2.

(b) Hence show that the force exerted on m_1 is equal and opposite to the force exerted on m_2.

(c) State any assumption you have made in answering parts (a) and (b).

Impulse

The impulse due to a force is the force times the length of time for which the force acts. Impulse is equal to the change in momentum.

$$\text{impulse} = Ft = (mv - mu)$$

The SI unit of impulse is the newton second, N s.

Impulse is often used to calculate the change in momentum when a force acts for a short time, for example when a bat hits a ball.

Some Higher questions include a graph showing how the force changes with time. For these questions impulse is equal to the area under the force–time (*F–t*) graph.

Exercises

Exercise 15 Impulse

1 A stationary golf ball of mass 48·0 g is struck by a putter. Immediately after the stroke the velocity of the golf ball is 0·380 m s^{-1}.

(a) Calculate the change in momentum of the golf ball.

(b) State the impulse on the ball.

(c) The time of contact between the ball and putter is 6·0 ms. Calculate the average force exerted by the putter on the ball.

2 The force–time graph for a stationary ball being hit by a baseball bat is shown below.

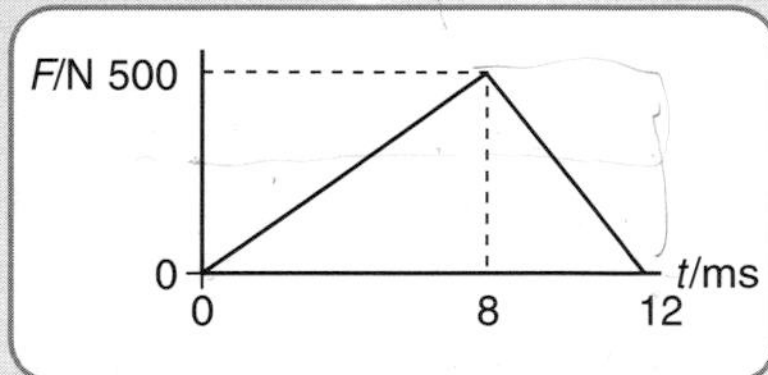

(a) Calculate the impulse on the ball.

(b) State the change in momentum of the ball.

(c) Calculate the average force acting on the ball.

1.8 Density

What You Should Know

For Higher Physics you need to be able to:

- understand and use the quantity density
- solve problems on density
- explain why the densities of liquids and solids are similar and why the densities of gases are much lower
- describe how to measure the density of air.

Density measures how tightly matter is packed in different materials. It is defined as the mass per unit volume. The symbol for density used by SQA is the Greek letter rho, ρ, and the SI unit for density is kg m^{-3}.

Learn the relationship $\rho = \dfrac{m}{V}$.

Questions and Answers

Calculate the mass of a cube of copper of side 20 mm.
The density of copper = $8{\cdot}96 \times 10^3$ kg m^{-3}.

$$\text{length of cube sides} = 20 \text{ mm} = 0{\cdot}020 \text{ m} \Rightarrow V = 0{\cdot}020 \times 0{\cdot}020 \times 0{\cdot}020 = 8{\cdot}0 \times 10^{-6} \text{ m}^3$$

Substitute in $\rho = \frac{m}{V}$

$$\Rightarrow \; 8{\cdot}96 \times 10^3 = \frac{m}{8{\cdot}0 \times 10^{-6}}$$

$$\Rightarrow \; m = 7{\cdot}168 \times 10^{-2} = 7{\cdot}2 \times 10^{-2} \text{ kg}.$$

The densities of common materials are listed on the data page at the start of the Higher Physics paper. Table 1.1 gives some of the values that may be included there.

Table 1.1 Densities of some common materials

Substance	*Density*/kg m^{-3}	*Substance*	*Density*/kg m^{-3}
Aluminium	$2{\cdot}70 \times 10^3$	Air	1·29
Copper	$8{\cdot}96 \times 10^3$	Hydrogen	$9{\cdot}0 \times 10^{-2}$
Ice	$9{\cdot}20 \times 10^2$	Oxygen	1·43
Sea water	$1{\cdot}02 \times 10^3$		
Water	$1{\cdot}00 \times 10^3$		

You will need some of these values for the questions in *Exercise 16 Density*. If you need a particular density value during your Higher examination, it will either be given in the question or it will be on the data page.

Look at the values in Table 1.1. The left hand columns give information on solids and liquids and the right hand columns give information on gases. Notice that the density values on the left hand side are very large. The gas densities are of the order of 1000 times smaller.

The huge difference between gases and both solids and liquids is due to differences in the spacing of the particles. In solids and liquids the particles are very close together. In gases the particles are about ten times further apart.

Other differences in density are due to differences in the masses of the particles and differences in the ways in which the particles are packed together.

Exercises

Exercise 16 Density

1 The length of the side of a cube of aluminium is 30·0 cm.
   (a) Calculate the mass of the cube.
   (b) Calculate the volume of sea water that has the same mass as the aluminium cube.
   (c) Calculate the volume of air that has the same mass as the aluminium cube.
   (d) Why is the answer to part (c) much greater than the answer to part (b)?

2 An ice sculpture has a mass of 350 kg.
   (a) Calculate the volume of ice in the sculpture.
   (b) The sculpture melts and all of the melt water is collected in a tank. Explain whether the volume of the melt water is less than, equal to or more than the volume of the ice.

Experiments

Measuring the density of air

To measure the density of any substance you need to measure the mass and the volume of a fixed sample of the material. To measure the density of air you need a very accurate balance and a special container such as a graduated syringe (see Figure 1.3).

The inside volume of the syringe is marked on a scale on the side of the syringe. The plunger of the syringe is a sealed glass cylinder so that when the plunger is pushed fully down there is no air inside the syringe.

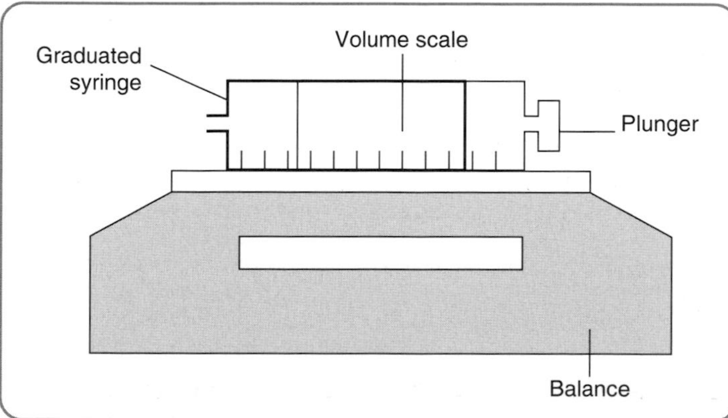

Figure 1.3

Use the balance to find the mass of the empty syringe. Pull back the plunger to one of the scale markings and note the volume of air inside the syringe. Weigh the syringe again and note the mass of the syringe plus air. Calculate the mass of air. Repeat for several different scale markings.

For each set of measurements divide the mass of air inside the syringe by the volume of this air.

Calculate the average value of the density of the air.

Exercises

Exercise 17 Measuring the density of air

1 An empty graduated syringe has a mass of 430·165 g. The plunger is pulled back until the volume of air in the syringe is 750 ml. The mass of the syringe plus air is 431·133 g.

(a) Calculate the density of air.

(b) The scale on the graduated syringe is marked from 0 to 1000 ml. Using the same equipment, state one change to the experimental procedure which would improve the accuracy of the measurement of the density of air.

1.9 Pressure

What You Should Know

For Higher Physics you need to be able to:

- understand and use the quantity pressure
- solve problems on pressure
- understand and use $P = \rho gh$ for pressure in fluids (liquids and gases)
- solve problems on pressure in fluids
- understand and use buoyancy force.

Pressure is the force per unit area acting perpendicular to a surface. The SI unit of pressure is the pascal (Pa). 1 Pa = 1 N m^{-2}. Pressure is a scalar quantity.

$$P = \frac{F}{A}$$

Questions and Answers

A cube of copper of side 100 mm rests on a horizontal table. Calculate the pressure exerted by the copper on the table surface.

$l = 100 \text{ mm} = 0{\cdot}100 \text{ m} \quad \Rightarrow \quad V = 0{\cdot}100 \times 0{\cdot}100 \times 0{\cdot}100 = 1{\cdot}00 \times 10^{-3} \text{ m}^3$ *volume*

$\text{and } A = 0{\cdot}100 \times 0{\cdot}100 = 1{\cdot}00 \times 10^{-2} \text{ m}^2$ *area of face*

$\rho_{copper} = 8{\cdot}96 \times 10^3 \text{ kg m}^{-3} \qquad \rho = \frac{m}{V}$ *value of ρ from Section 1.8, Table 1.1*

$$\Rightarrow 8{\cdot}96 \times 10^3 = \frac{m}{1{\cdot}00 \times 10^{-3}}$$

$$\Rightarrow \quad m = 8{\cdot}96 \text{ kg}$$

$g = 9{\cdot}8 \text{ N kg}^{-1} \quad \Rightarrow \quad mg = 87{\cdot}8 \text{ N}$ *weight = downward force*

$$P = \frac{F}{A}$$

$$= \frac{87{\cdot}8}{1{\cdot}00 \times 10^{-2}}$$

$$= 8780 \text{ Pa} = 8800 \text{ Pa}$$

Exercises

Exercise 18 Pressure

1 A stationary car has four wheels. The area of each tyre in contact with the ground is $0{\cdot}032 \text{ m}^2$. The average pressure exerted on the ground by the tyres is $9{\cdot}2 \times 10^4$ Pa. Calculate the mass of the car.

2 A woman of mass 46 kg walks across a wooden floor. The area of the heel on each of the woman's shoes is $0{\cdot}64 \text{ cm}^2$. Calculate the pressure acting on the floor when all of the woman's weight is on one heel.

Pressure in fluids

A fluid is any material which flows – all liquids and gases are fluids.

Pressure at a point in a fluid is due to the weight of the fluid above that point. This depends on the depth, the density and the gravitational field strength.

In a fluid $P = g\rho h$.

Pressure at a point in a fluid acts in all directions. This means that the pressure could result in a force in any direction. The direction of the force is perpendicular to any surface placed in the fluid.

When large objects, for example submarines, are completely submerged in a liquid the pressure on the lower side is greater than the pressure on the upper side. This pressure difference causes an upward force called **buoyancy force**.

In the Higher examination, you must use the full term 'buoyancy force' when you name this force. You can also call this force **upthrust**.

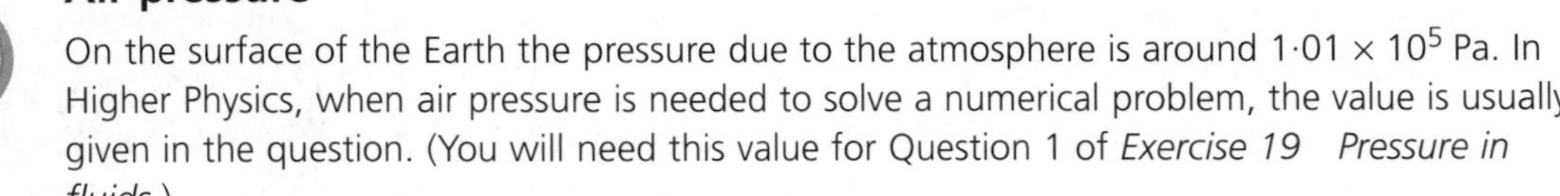

Air pressure

On the surface of the Earth the pressure due to the atmosphere is around $1{\cdot}01 \times 10^5$ Pa. In Higher Physics, when air pressure is needed to solve a numerical problem, the value is usually given in the question. (You will need this value for Question 1 of *Exercise 19 Pressure in fluids*.)

When you are calculating the pressure at a given depth in a lake or the sea you must remember to add the pressure due to the Earth's atmosphere to the pressure due to the water.

Air pressure changes with the weather and with height above the surface of the Earth. You may need to use these ideas in explanations or when you are evaluating experimental work.

Questions and Answers

A rectangular piece of glass of area $0{\cdot}32$ m^2 is placed horizontally at a depth of 15 m in a tank of sea water.

(a) Calculate the pressure acting on the surface of the glass.

(b) Calculate the downward force acting on the upper side of the glass.

(c) State the upward force acting on the lower side of the glass.

(d) What assumption have you made to get your answer to part (c)?

(a)
$h = 15$ m
$g = 9{\cdot}8$ m s^{-2}
$\rho = 1{\cdot}02 \times 10^3$ kg m^{-3}
air pressure $= 1{\cdot}01 \times 10^5$ Pa

$$P = g\rho h + \text{air pressure}$$
$$= (9{\cdot}8 \times 1{\cdot}02 \times 10^3 \times 15) + 1{\cdot}01 \times 10^5$$
$$= 2{\cdot}509 \times 10^5 = 2{\cdot}5 \times 10^5 \text{ Pa}$$

(b) $A = 0{\cdot}32$ m^2

$$P = \frac{F}{A}$$

$$\Rightarrow \quad 2{\cdot}5 \times 10^5 = \frac{E}{0{\cdot}32}$$

$$\Rightarrow \quad F = 8{\cdot}0 \times 10^4 \text{ N}$$

(c) Upward force on underside of the glass = $8{\cdot}0 \times 10^4$ N.

(d) This assumes that the glass is thin $\Rightarrow$ there is a negligible difference between the pressure on the lower side and the pressure on the upper side.

Exercises

Exercise 19 Pressure in fluids

1 A diving bell has a vertical rectangular window that is 0·30 m high and 0·40 m wide. The diving bell is suspended in fresh water so that the centre of the window is at a depth of 25 m below the surface.

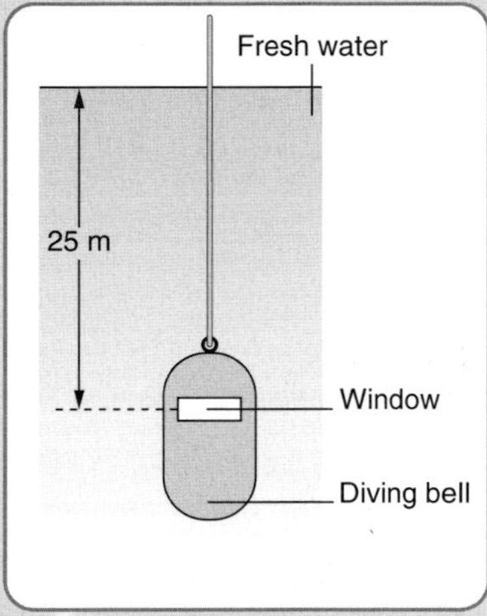

(a) Calculate the pressure acting at the centre of the window.

(b) (i) Calculate the horizontal force acting on the window.

(ii) State any assumption you have made.

2 A small metal block is suspended from a spring balance at a depth *h* below the surface of a fresh water lake. The reading on the balance is noted. The cube is then suspended at the same depth below the surface of the sea and the reading on the balance is again noted.

(a) Explain how the readings on the balance compare with the weight of the metal block. (Values are not required.)

(b) How do the two balance readings compare with each other? (Values are not required.)

(c) Explain your answer to part (b).

(d) The experiment is repeated with the metal block suspended at a depth ½ *h* below the surface of both the lake and the sea. Explain any difference in the readings compared with those in the original experiment.

3 Why is the air pressure at the top of a mountain less than the air pressure at sea level?

1.10 Kelvin temperature and the gas laws

What You Should Know

For Higher Physics you need to be able to:

- convert temperatures between kelvins and degrees celsius
- understand and use the pressure–temperature, temperature–volume and volume–pressure laws for a fixed mass of gas
- understand and use the general gas equation
- explain the concept of absolute zero of temperature.

Kelvin temperature scale

The zero of the kelvin temperature scale is – 273·15 °C. For Higher Physics you should use the approximate value – 273 °C. This means that 0 °C = + 273 K.

When you are changing degrees celsius (centigrade) to kelvins and back remember that temperatures on the kelvin scale are much bigger numbers. Also on the kelvin scale temperatures are never negative.

One celsius degree is exactly the same size as one kelvin. *Changes* in temperature are exactly the same in both celsius degrees and kelvins.

Exercises

Exercise 20 Kelvin temperature scale

1 Convert the following temperatures to kelvins.
   (a) 100 °C
   (b) – 100 °C
   (c) 273 °C
   (d) – 270 °C
   (e) 27 °C.

2 Convert the following temperatures to degrees celsius.
   (a) 100 K
   (b) 400 K
   (c) 270 K
   (d) 73 K
   (e) 350 K.

3 A metal block is heated from 20 °C to 350 °C.
   (a) Calculate the temperature change in °C.
   (b) State the temperature change in kelvins.

Absolute zero

The zero of the kelvin scale is the lowest possible temperature. It is not possible to get any temperature lower than 0 K. For this reason 0 K is called the **absolute zero** of temperature.

Gas laws

There are three gas laws.

1 The pressure of a fixed mass of gas at constant temperature is inversely proportional to its volume.

$$PV = \text{constant} \quad \textbf{or} \quad P_1V_1 = P_2V_2$$

2 The pressure of a fixed mass of gas at constant volume is directly proportional to its temperature measured *in kelvins*.

$$\frac{P}{T} = \text{constant} \quad \textbf{or} \quad \frac{P_1}{T_1} = \frac{P_2}{T_2}$$

3 The volume of a fixed mass of gas at constant pressure is directly proportional to its temperature measured *in kelvins*.

$$\frac{V}{T} = \text{constant} \quad \textbf{or} \quad \frac{V_1}{T_1} = \frac{V_2}{T_2}$$

Learn these as you could be asked to state any one of them.

The three gas law relationships have been combined into the following general gas equation.

$$\frac{PV}{T} = \text{constant} \quad \textbf{or} \quad \frac{P_1V_1}{T_1} = \frac{P_2V_2}{T_2}$$

In Higher Physics you have to be able to solve problems using any of the gas law relationships. It is easiest to use the general gas equation all the time – this way you only have to practise using one equation. It is also a good way of ensuring you do not miss out any data.

Questions and Answers

A helium balloon has a volume of 0·14 m^3 at ground level where the temperature is 17 °C and the air pressure is 1·01 × 10^5 Pa. The balloon is released and rises. The balloon bursts when its volume is double its original volume. At this point the air temperature is 7 °C.

(a) Calculate the pressure of the helium when the balloon bursts.

(b) State any assumption you have made in your calculation.

(a)

$V_1 = 0{\cdot}14\ \text{m}^3$ $\qquad \dfrac{P_1V_1}{T_1} = \dfrac{P_2V_2}{T_2}$

$P_1 = 1{\cdot}01 \times 10^5\ \text{Pa}$ $\qquad \Rightarrow \dfrac{1{\cdot}01 \times 10^5 \times 0{\cdot}14}{290} = \dfrac{P_2 \times 0{\cdot}28}{280}$

$T_1 = 17\ °\text{C} = 290\ \text{K}$ $\qquad \Rightarrow P_2 = 4{\cdot}876 \times 10^4$

$V_2 = 0{\cdot}28\ \text{m}^3$ $\qquad = 4{\cdot}9 \times 10^4\ \text{Pa}$

$P_2 = ?$

$T_2 = 7\ °\text{C} = 280\ \text{K}$

(b) This assumes that the temperature of the helium is equal to the air temperature.

Exercises

Exercise 21 Gas laws

1 Gas trapped in the cylinder of a graduated syringe is at a pressure of $8{\cdot}17 \times 10^4$ Pa. The gas is compressed until its volume is one-third of its original volume.

(a) Assuming the temperature of the gas stays constant calculate the final pressure of the gas.

(b) In practice, the temperature of the gas rises as it is compressed. Explain the effect, if any, that this has on the final pressure of the gas.

2 A diving bell of internal volume 20 m^3 is lowered from just above the surface of the sea until the water surface inside the bell is at a depth of 32 m below the surface of the sea.

air pressure = $1{\cdot}01 \times 10^5$ Pa

density of sea water = $1{\cdot}02 \times 10^3$ kg m^{-3}

(a) Calculate the pressure of the air inside the lowered diving bell.

(b) Hence find the volume of the air inside the lowered diving bell.

3 A student uses the apparatus below to investigate the relationship between the pressure and temperature of a fixed mass of gas.

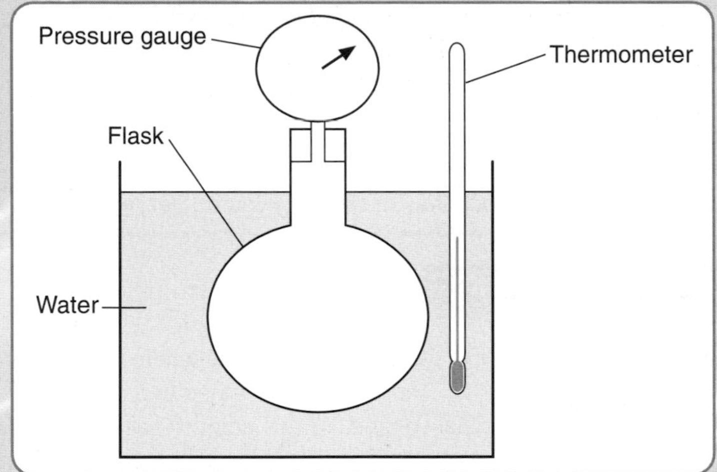

The water in the flask is heated gradually. Temperature and pressure readings are taken at regular intervals. The student obtains the following results.

Temperature/°C	20	30	40	50	60	70
Pressure/x 10^5 Pa	1·010	1·044	1·079	1·113	1·148	1·182

Use all of the results to establish the relationship between the pressure and temperature of the gas.

1.11 The kinetic model of matter

What You Should Know

For Higher Physics you need to be able to:

- use the kinetic model to explain pressure of a gas
- explain the pressure–temperature, temperature–volume and volume–pressure laws in terms of the kinetic model.

Kinetic model of matter

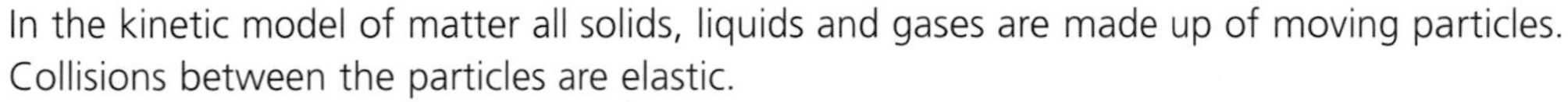

In the kinetic model of matter all solids, liquids and gases are made up of moving particles. Collisions between the particles are elastic.

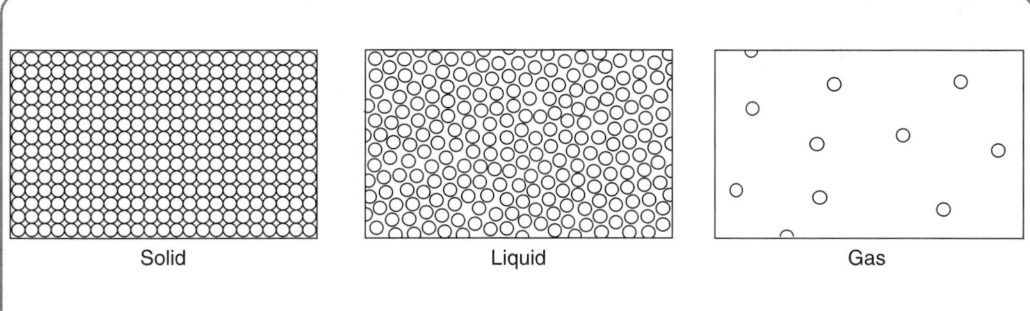

Figure 1.4

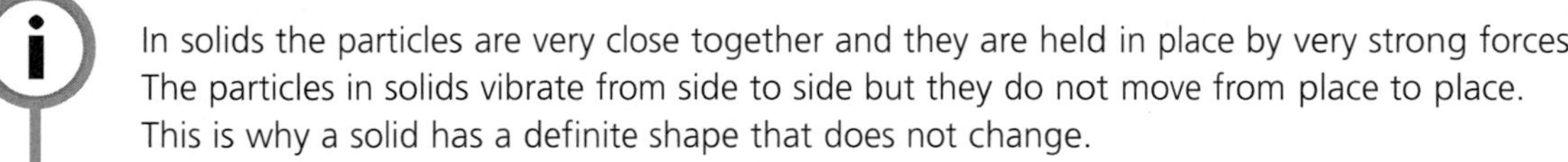

In solids the particles are very close together and they are held in place by very strong forces. The particles in solids vibrate from side to side but they do not move from place to place. This is why a solid has a definite shape that does not change.

In liquids the particles are very close together and they are held together by forces that are weaker than the forces in solids. The particles in liquids move about randomly. This is why a liquid flows and takes on the shape of any container in which it is placed.

In gases the particles are about ten times further apart than the particles in solids and liquids. The forces between gas particles are very weak. This is why gases expand to fill any container in which they are placed.

When temperature rises the average kinetic energy of particles increases.

Pressure of gases

Gas pressure on a surface is due to gas particles colliding with the surface. The pressure produced by these collisions depends on the following:

- the force with which particles hit the surface – this depends on the mass and the speed of the particles
- the number of particle collisions with the surface per second.

In Higher Physics, if you are asked to explain gas pressure, you must refer to the **number of collisions per second**. If you only mention the number of collisions you will lose marks.

Gas laws

Learn the following explanations and make sure you cover all of the important points.

Pressure–volume at constant temperature: As the temperature is constant the average kinetic energy of the particles does not change and so the particles hit the container wall with the same average force. When the volume of the gas is increased the particles have to travel further between collisions with the container walls. There are fewer particle collisions per second with the walls. This causes the pressure to fall.

Pressure–temperature at constant volume: When temperature rises the average kinetic energy of particles increases and the particles move faster. The particles hit the walls with greater force on average. Providing the volume does not change the particles also hit the walls more often each second. These two effects cause the pressure to increase.

Volume–temperature at constant pressure: As temperature falls the kinetic energy of the particles decreases and the particles move more slowly. The particles hit the wall with less force on average. This tends to make the pressure decrease. The volume of the gas decreases so that the particles hit the walls more often each second. This tends to increase the pressure. The two effects cancel each other out and the pressure stays constant.

Exercises

Exercise 22 Kinetic theory

1 Ice floats on water because the density of ice at 0 °C is less than the density of water at 0 °C. Explain this difference in density in terms of the kinetic model of matter.

2 At the end of a long journey, the pressure inside a car tyre is higher than it is at the start of the journey. Explain this effect in terms of the kinetic model.

3 (a) Explain why the volume of a weather balloon increases as the balloon rises through the Earth's atmosphere.

(b) Explain what happens in terms of the kinetic model.

Chapter 2

ELECTRICITY

2.1 Electric fields

What You Should Know

For Higher Physics you need to be able to:

- understand and use the term electric field
- define potential difference
- understand and use the relationship $W = QV$
- solve problems on charge moving in electric fields.

An electric field is a region of space where charge experiences an electrical force. Learn this definition as you could be asked to state it.

The direction of an electric field is the direction of the force experienced by a positive charge in the field. The force experienced by a negative charge is in the opposite direction to the direction of the electric field.

Figure 2.1 shows the shapes of two electrical fields.

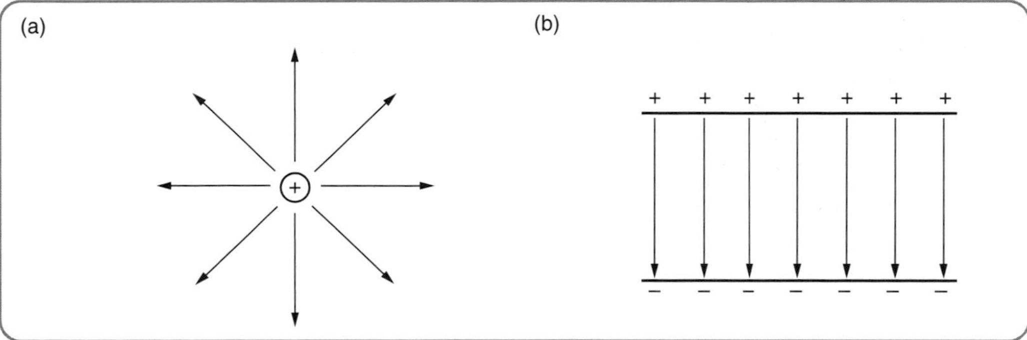

Figure 2.1 (a) Electric field around a positive point charge, (b) electric field between charged plates

You do *not* have to be able to draw the shapes of these fields for Higher Physics but understanding the electric field between charged plates will help you understand capacitors.

In an electrical circuit the electrical supply applies an electric field to the conductors. This causes free electric charges in the conductors to move. In Higher Physics the free charges in circuits are usually electrons.

Charge, work and potential difference

When an electrical force causes charge to move, the work done is equal to the charge times the potential difference through which the charge is moved.

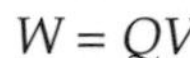

$$W = QV$$

When Q = 1 C, $V = W$, i.e. the potential difference between two points is equal to the work done in moving 1 C of charge between the points. This is the definition of potential difference and you could be asked to state it.

Also, when 1 J of work is done in moving 1 C of charge from one point to another, the potential difference between the points is 1 V, i.e. 1 volt = 1 joule per coulomb.

When charge is moved in an electric field the work done depends only on the start and finish points. It does not depend on the path followed by the charge.

Exercises

Exercise 23 Electric fields

1 The work done in moving 1 C of charge around an electric circuit is 15 J. State the potential difference between the terminals of the supply.

2 Calculate the work required to move 40 mC through a potential difference of 1·8 kV.

3 Points A, B, C and D are in an electric field. A charge of 0·30 C is moved from point A to point B and then to point C.

The total work done is 1·2 J.

(a) Calculate the potential difference between A and C.

(b) A second charge of 0·30 C is moved from point A to point D and then to point C. State the total work done in moving this charge. Justify your answer.

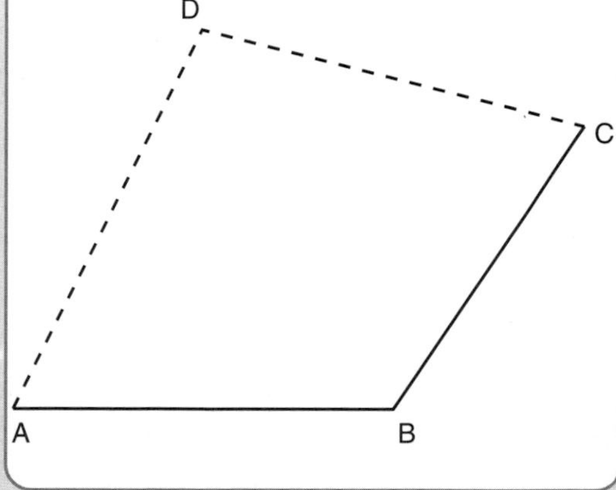

4 In a TV tube, electrons are accelerated through a potential difference of 2·30 kV.

(a) Calculate the kinetic energy gained by the electrons.

(b) Hence calculate the final speed of the electrons. State any assumption you have made in your calculation.

2.2 e.m.f. and internal resistance

What You Should Know

For Higher Physics you need to be able to:

- understand and use the quantities e.m.f., internal resistance and terminal p.d.
- use conservation of energy to explain that the sum of the e.m.f.s around a closed circuit is equal to the sum of the p.d.s around the circuit
- solve problems on e.m.f. and internal resistance
- describe how to measure e.m.f. and internal resistance.

The e.m.f. of a source of electrical energy is the electrical potential energy supplied to each coulomb of charge which passes through the source.

A source of electrical energy is equivalent to an e.m.f. in series with a small resistor. This small resistor is called the internal resistance. SQA uses the symbol E for e.m.f. and the symbol r for internal resistance.

The circuit diagram in Figure 2.2 shows a battery of e.m.f. E and internal resistance r connected to a load resistor R. The current in the circuit is I.

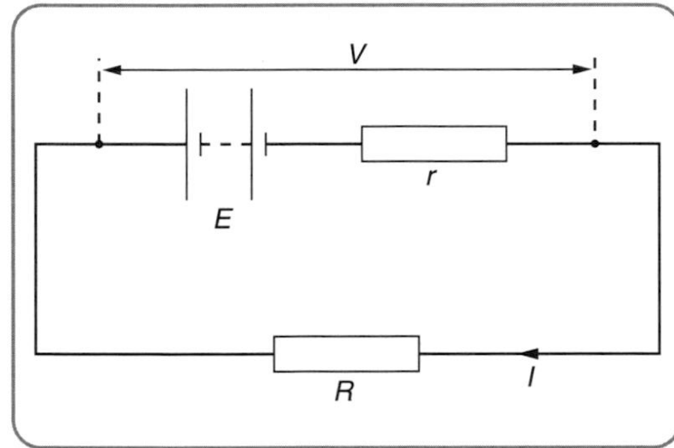

Figure 2.2

The potential difference across the battery terminals, i.e. terminal p.d., is given by

$V = E - Ir$

$\Rightarrow E = V + Ir.$ Equation ①

V is also the p.d. across the load resistor $\Rightarrow V = IR.$

Substitute for V in Equation ① $\Rightarrow E = IR + Ir = I(R + r).$

Learn these relationships for e.m.f. – this will save you time in the examination.

The term 'Ir' is called the 'lost volts'.

Energy in circuits

When charge Q moves round a circuit it gains electrical potential energy as it passes through an e.m.f. The energy gained = $Q \times$ e.m.f. When there is more than one e.m.f. the

$$\text{total energy gained} = Q \times (\text{sum of the e.m.f.s}).$$

When the charge passes through a resistor the electrical potential energy is converted to other forms of energy. The electrical energy lost = $Q \times$ p.d. When there is more than one p.d.

$$\text{total energy lost} = Q \times (\text{sum of the p.d.s}).$$

As energy is conserved, total energy gained = total energy lost

$$\Rightarrow \text{sum of the e.m.f.s} = \text{sum of the p.d.s}$$

$$\Sigma E = \Sigma IR.$$

The symbol Σ means 'sum'.

Power

Electrical power is electrical work done per second $\Rightarrow P = \frac{W}{T} = \frac{QV}{t} = \frac{Q}{t}V$, i.e. $P = IV$.

Substituting $V = IR$ in this equation $\Rightarrow P = IV = I \times IR = I^2R$.

Similarly substituting for $I \Rightarrow P = IV = \frac{V}{R} \times V = \frac{V^2}{R}$.

Learn these relationships. Choosing the right version can save you time during your Higher Physics exam.

Exercises

Exercise 24 e.m.f. and internal resistance

1 An electrical supply of e.m.f. 20 V and internal resistance 2·0 Ω is connected to a load resistor of 38 Ω.

(a) Calculate the current drawn from the supply.

(b) Calculate the power dissipated in the load resistor.

(c) Calculate the power wasted.

2 A battery of e.m.f. 12 V is connected in series with a load resistor of 23 Ω. The current in the circuit is 0·50 A. Calculate the internal resistance of the battery.

3 (a) A student states that the e.m.f. is the maximum possible terminal p.d. of a source of electrical energy. Explain whether the student is correct.

(b) The current drawn from a source of electrical energy is gradually increased from zero. Describe what happens to the terminal p.d. of the source.

Experiments

Measuring e.m.f. and internal resistance

The circuit shown in Figure 2.3 can be used to measure the e.m.f. and internal resistance of an electrical supply.

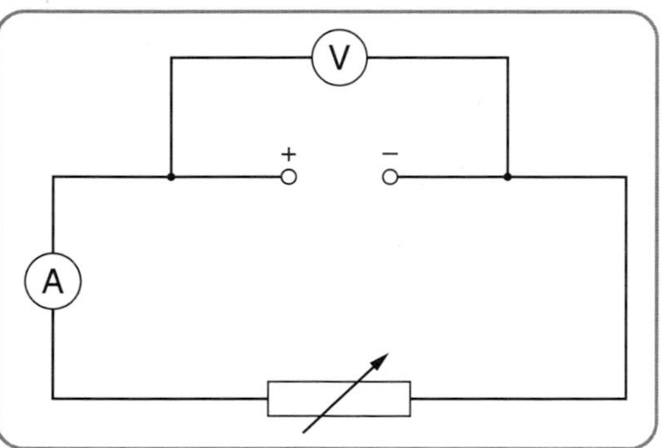

Figure 2.3

The voltmeter measures the terminal p.d. and the ammeter measures the current drawn from the supply. Close the switch and note the voltmeter and ammeter readings. Adjust the variable resistor and again note the meter readings. Repeat until at least five pairs of readings have been obtained.

Plot a graph of current (x-axis) against terminal p.d. (y-axis).

A straight line graph with a negative gradient is obtained.

Now $V = E - Ir$. This equation can be re-written as $V = -rI + E$.

Compare this with the general equation of a straight line $y = mx + c$.

The way the graph is drawn, V is equivalent to y and I is equivalent to x

$\Rightarrow E$ is equivalent to c, the y-intercept (the value of V when $I = 0$)

and $\Rightarrow -r$ is equivalent to m, the gradient.

Exercises

Exercise 25 Measuring e.m.f. and internal resistance

1 A student sets up the apparatus as shown in Figure 2.3 to measure the e.m.f. and resistance of a d.c. supply.

The student carries out the experiment and obtains the following results.

Current/A	0·40	0·60	0·80	1·00	1·20	1·40	1·60
Terminal p.d./V	8·60	8·41	8·19	8·02	7·78	7·59	7·40

Use all of the results to find the

(a) e.m.f. of the supply

(b) internal resistance of the supply.

2.3 Resistors in circuits

What You Should Know

For Higher Physics you need to be able to:

- use conservation of energy to derive the relationship for the total resistance of a number of resistors in series
- use conservation of charge to derive the relationship for the total resistance of a number of resistors in parallel
- solve problems on electrical circuits containing resistors.

You could be asked to derive either of the following relationships so learn them.

Resistors in series

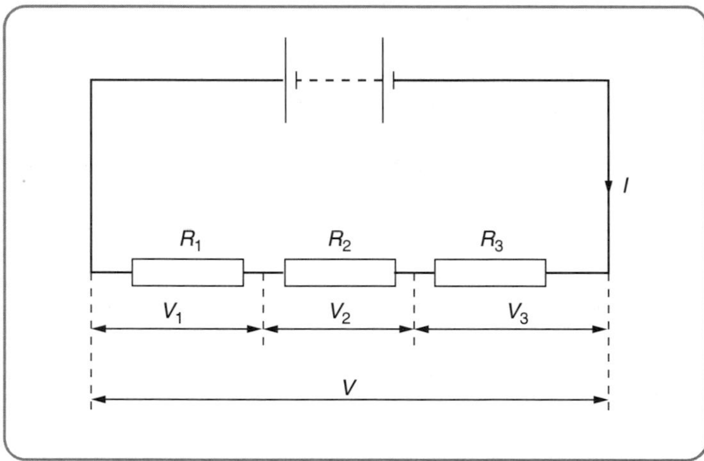

Figure 2.4

In a series circuit there is only one path for electrons to follow so the current is the same at all points. Charge going round the circuit gains electrical potential energy as it goes through the battery. It loses this energy as it goes through the resistors.

Consider charge Q going round the circuit. The energy gained = QV.

The p.d. across $R_1 = IR_1 \Rightarrow$ the energy lost in $R_1 = Q \times IR_1$.

Similarly the energy lost in $R_2 = Q \times IR_2$

and the energy lost in $R_3 = Q \times IR_3$.

As energy is conserved total energy gained = total energy lost

$$\Rightarrow \quad QV = QIR_1 + QIR_2 + QIR_3$$

$$\Rightarrow \quad V = IR_1 + IR_2 + IR_3$$

$$\Rightarrow \quad V = I(R_1 + R_2 + R_3). \qquad \text{Equation ①}$$

If the total resistance of the circuit is R, then $V = IR$. Equation ②

Comparing Equation ① and Equation ② ⇒

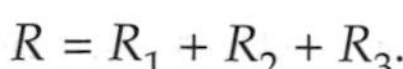

$$R = R_1 + R_2 + R_3.$$

Resistors in parallel

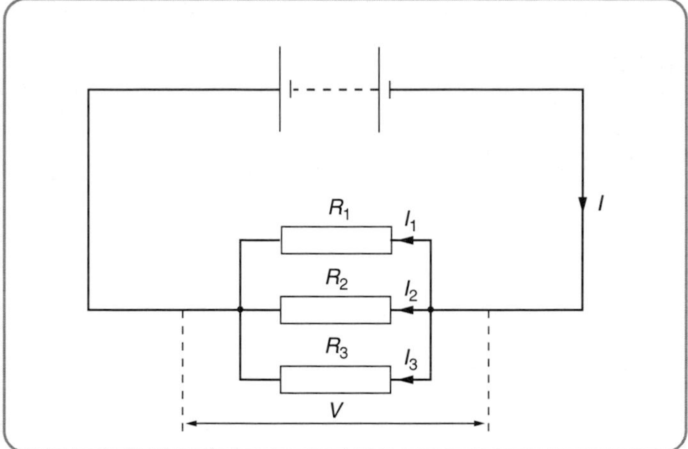

Figure 2.5

In a parallel circuit, the battery is connected directly to each resistor so the p.d. across each resistor is V. The current in the battery is I.

The currents in R_1, R_2 and R_3 are I_1, I_2 and I_3 respectively.

As charge is conserved

$$I = I_1 + I_2 + I_3. \qquad \text{Equation ①}$$

If the total resistance of the circuit is R, then $V = IR \Rightarrow I = \frac{V}{R}$.

Similarly $V = I_1R_1 \Rightarrow I_1 = \frac{V}{R_1}$ **and** $V = I_2R_2 \Rightarrow I_2 = \frac{V}{R_2}$ **and** $V = I_3R_3 \Rightarrow I_3 = \frac{V}{R_3}$.

Substituting in Equation ① $\Rightarrow \frac{V}{R} = \frac{V}{R_1} + \frac{V}{R_2} + \frac{V}{R_3}$

$$\Rightarrow \frac{1}{R} = \frac{1}{R_1} + \frac{1}{R_2} + \frac{1}{R_3}.$$

When you use this relationship, after you work out the sum you *must* invert to get R. Not inverting is a very common mistake in Higher Physics questions – do not let it happen to you.

Basic circuit rules

Make sure you really understand the basic circuit rules for current and voltage.

Series:	**current** is the **same** at all points
	p.d.s add up to the p.d. across the battery
Parallel:	**p.d.** is the **same** across all components
	currents add up to the current in the battery

Sometimes in a Higher Physics question you might think that there is not enough information. If you get a question like this take one step at a time. The first step often involves applying one of the basic circuit rules. The solutions to questions 4 and 5 of *Exercise 26 Series and parallel circuits* show how much you can work out from just a few pieces of data.

Exercises

Exercise 26 Series and parallel circuits

1 Each coulomb of charge gains 20 J of electrical potential energy as it passes through a supply. State the p.d. across the terminals of the supply.

2 **Series circuit**: A student sets up the following circuit.

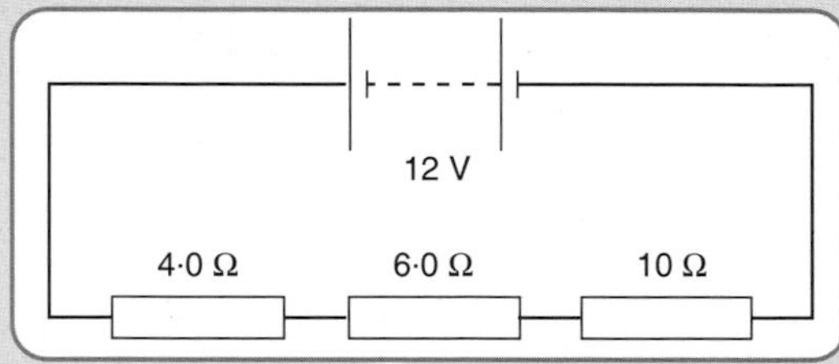

(a) Calculate the current drawn from the battery.

(b) Calculate the p.d. across each resistor.

(c) Compare the sum of the p.d.s across the resistors with the p.d. across the battery terminals.

3 **Parallel circuit**: A student sets up the following circuit.

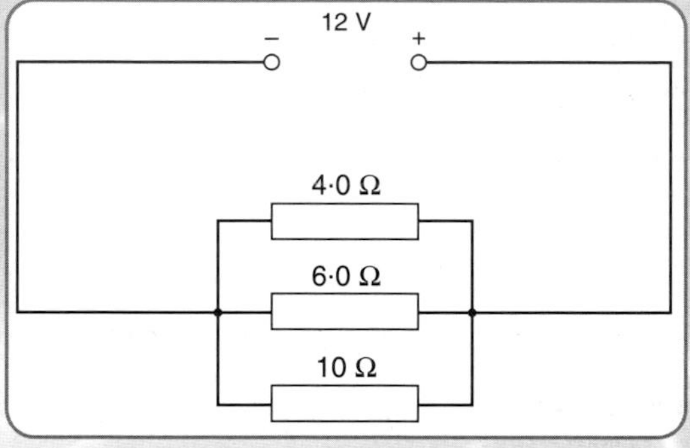

(a) Calculate the current drawn from the battery.

(b) Calculate the current in each resistor.

(c) Compare the sum of the currents in the resistors with the current drawn from the battery.

4 In the following circuit the current in the 4·0 Ω resistor is 0·30 A.

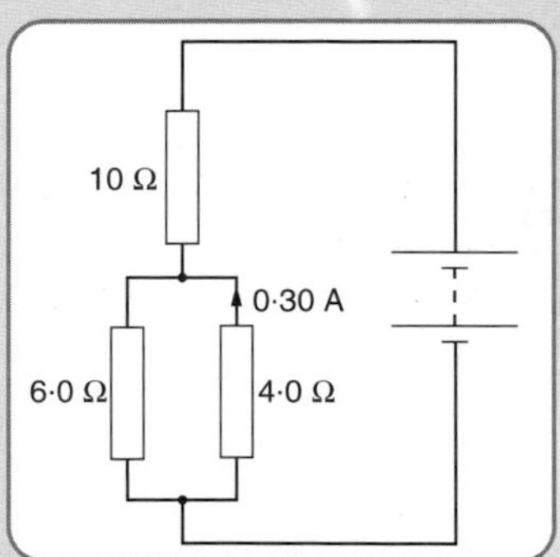

Exercises continued

Exercises continued

(a) Calculate the p.d. across the parallel branches.

(b) Calculate the current in the 6·0 Ω resistor.

(c) Calculate the current in the 10 Ω resistor.

(d) Calculate the p.d. across the battery terminals.

5 In the following circuit the current in the 4·0 kΩ resistor is 0·30 mA. The current in the high resistance voltmeter is negligible.

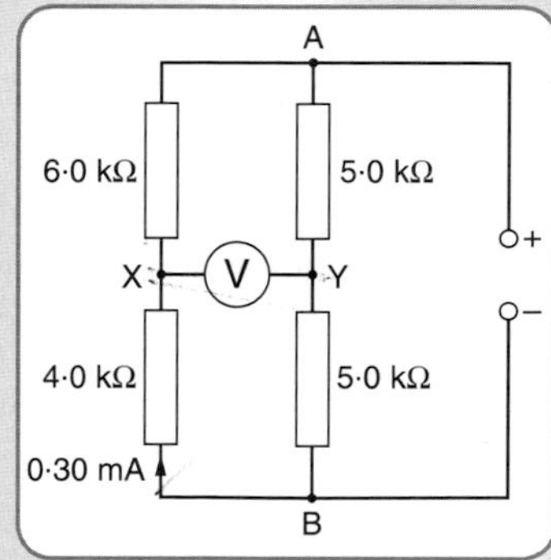

(a) Calculate the p.d. across the battery.

(b) Calculate the reading on the voltmeter (i.e. the p.d. between X and Y).

2.4 The Wheatstone bridge

What You Should Know

For Higher Physics you need to be able to:

- state and use the relationship for a balanced Wheatstone bridge
- state and use the relationship for an out-of-balance Wheatstone bridge
- solve problems on electrical circuits containing Wheatstone bridges.

The potential divider

The potential divider is not an explicit requirement of the Higher Physics course. However, understanding potential dividers will help you understand the Wheatstone bridge. Also, potential dividers are very common in questions on electronic circuits.

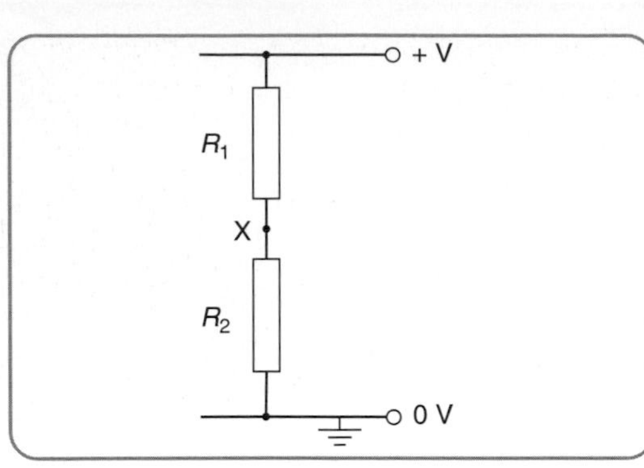

A potential divider usually consists of two resistors in series with a d.c. power supply as shown in Figure 2.6.

Figure 2.6

The values of the resistors R_1 and R_2 are chosen to control the voltage at point X.

The total resistance = $R_1 + R_2$.

Using $V = IR \Rightarrow$ current in the divider $= V \times \dfrac{1}{R_1 + R_2}$.

Using $V = IR$ again $\Rightarrow$ p.d. across $R_1 = V \times \dfrac{R_1}{R_1 + R_2}$.

Similarly the p.d. across $R_2 = V \times \dfrac{R_2}{R_1 + R_2}$.

These relationships are very useful for calculating p.d.s in series circuits.

The Wheatstone bridge

A Wheatstone bridge has four resistors arranged as shown in Figure 2.7 (it is like two potential dividers connected in parallel).

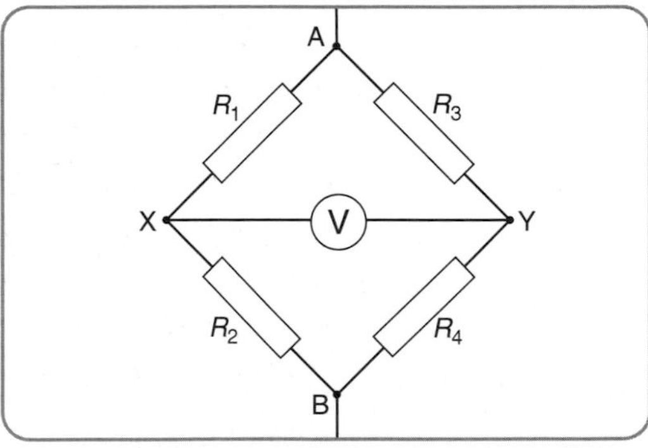

Figure 2.7

Electrically this is exactly the same as the arrangement of resistors in question 5 of *Exercise 26 Series and parallel circuits*. The circuit is just drawn differently. A Wheatstone bridge circuit could be drawn either way in your Higher Physics examination so make sure you recognise both formats.

- A Wheatstone bridge is balanced when the p.d. between X and Y is 0 V.
- For a **balanced** Wheatstone bridge $\dfrac{R_1}{R_2} = \dfrac{R_3}{R_4}$.
- When the bridge is balanced the current in XY is 0 A.

For an initially balanced Wheatstone bridge, when the value of one of the resistors is changed, the bridge goes **out of balance**. Providing the change in resistance is small, the out-of-balance p.d. across XY is directly proportional to the **change** in the resistance.

Higher Physics questions often include Wheatstone bridge circuits in which one of the resistors is a resistive sensor such as a light-dependent resistor (LDR) or a thermistor. Changes in the resistance of the sensor are used to control an electronic circuit.

Exercises

Exercise 27 Wheatstone bridge

1 In a Wheatstone bridge circuit, $R_1 = 20\ \Omega$, $R_2 = 60\ \Omega$ and $R_3 = 18\ \Omega$. Calculate the resistance of R_4 when the bridge is balanced.

2 A Wheatstone bridge circuit is set up as shown below.

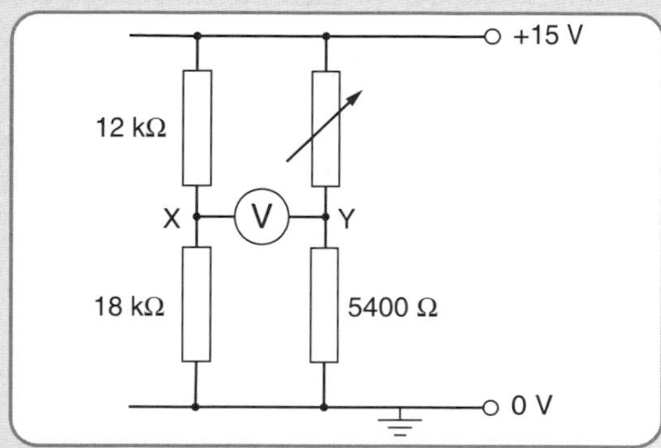

A high resistance voltmeter is connected between points X and Y.

(a) Calculate the p.d. across the 12 kΩ resistor. State any assumption you make in your calculation

(b) Calculate the p.d. across the 18 kΩ resistor.

(c) The resistance of the variable resistor is set at 2700 Ω. Calculate the reading on the voltmeter.

(d) Calculate the resistance of the variable resistor for which the voltmeter reading is zero.

(e) The zero of the voltmeter is at the centre of its scale. The resistance of the variable resistor is increased from 2700 Ω to 4000 Ω. Describe and explain how the reading on the voltmeter changes during this process.

3 The resistance of the thermistor in the circuit below is 3600 Ω at a temperature of 15 °C.

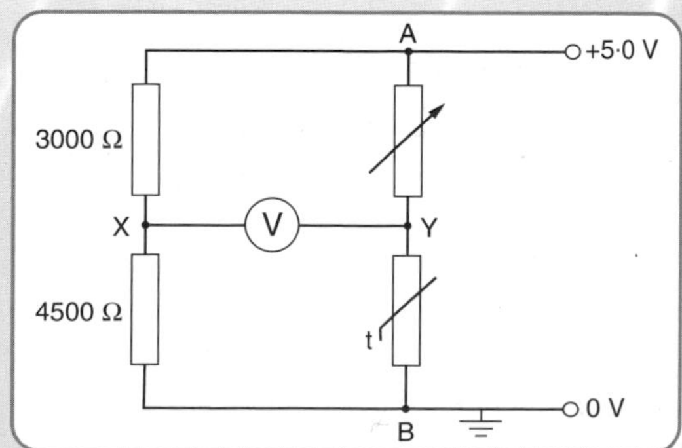

Exercises continued ➢

Exercises continued

(a) Calculate the value of the variable resistor for which the Wheatstone bridge is balanced at 15 °C.

(b) With the variable resistor at this setting, the circuit is placed in a container in which the temperature is initially 10 °C. The temperature inside the container is gradually increased to 20 °C. Explain what happens to the p.d. between points X and Y. Values are not required but you must state clearly which point, X or Y, is at a higher voltage at each stage of your explanation.

2.5 Alternating current and voltage

What You Should Know

For Higher Physics you need to be able to:

- state and use the relationships between peak current and r.m.s. current and between peak voltage and r.m.s. voltage
- state that in an a.c. circuit containing only resistors, current does not change when frequency changes
- solve problems on current and voltage in a.c. circuits
- describe how to measure a.c. frequency.

In Higher Physics the current–time and voltage–time graphs for a.c. supplies have the same shape as a sine wave, as shown in Figure 2.8.

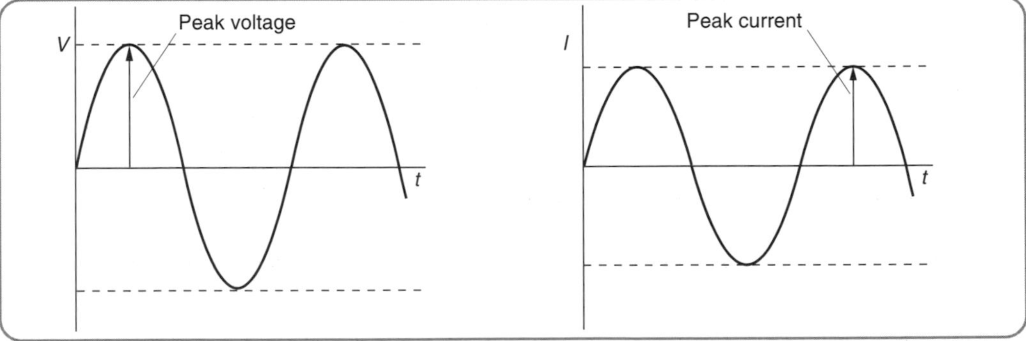

Figure 2.8

The peak values of current and voltage are the maximum values. The Higher Physics paper sometimes includes questions on the measurement of peak values using an oscilloscope.

The root mean square (r.m.s.) values are the 'effective' values, i.e. they are equal to d.c. values that produce the same effect in a resistor. Peak values are always bigger than r.m.s. values.

The relationships between peak values and r.m.s. values of voltage and current are

$$V_{\text{peak}} = \sqrt{2}\,V_{\text{r.m.s.}} \quad \textbf{and} \quad I_{\text{peak}} = \sqrt{2}\,I_{\text{r.m.s.}}.$$

Both the peak and r.m.s. values are used in some circuit relationships. For example

$$V_{\text{peak}} = I_{\text{peak}}R \quad \text{and} \quad V_{\text{r.m.s.}} = I_{\text{r.m.s.}}R.$$

For electrical energy calculations you must use r.m.s. values.

The frequency of an a.c. supply is the number of complete cycles of voltage (or current) per second. The above relationships apply to any a.c. frequency.

In a circuit containing only resistors, current is independent of frequency; when frequency is changed, current remains the same.

Experiments

Measuring a.c. frequency

To measure the frequency of an a.c. supply, connect the supply to the input terminals of a calibrated oscilloscope, as shown in Figure 2.9.

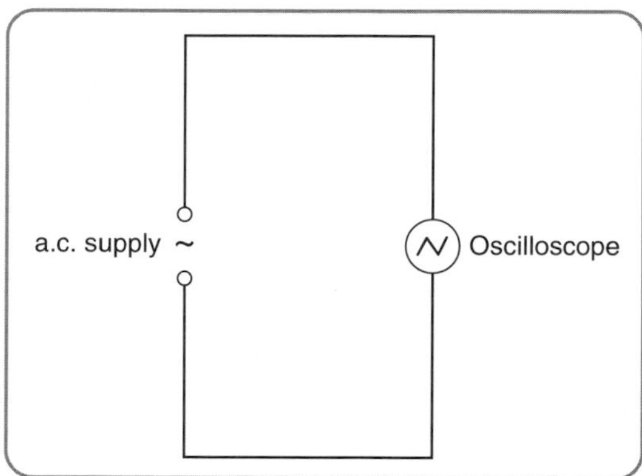

Figure 2.9

Adjust the time-base of the oscilloscope to give a suitable number of waves on the screen. Work out the number of divisions on the oscilloscope for one complete wave and multiply this by the time-base setting. The number obtained is the period T of the a.c. supply.

Use the relationship $f = \frac{1}{T}$ to find the frequency of the supply.

Time-base settings are often given in odd units such as ms cm^{-1} so be very careful with units.

Exercises

Exercise 28 Alternating current and voltage

1 The peak value of the voltage of an a.c. supply is 140 V. Calculate the r.m.s. voltage of this supply.

2 A 7·0 V r.m.s. a.c. supply is connected as shown in the circuit below.

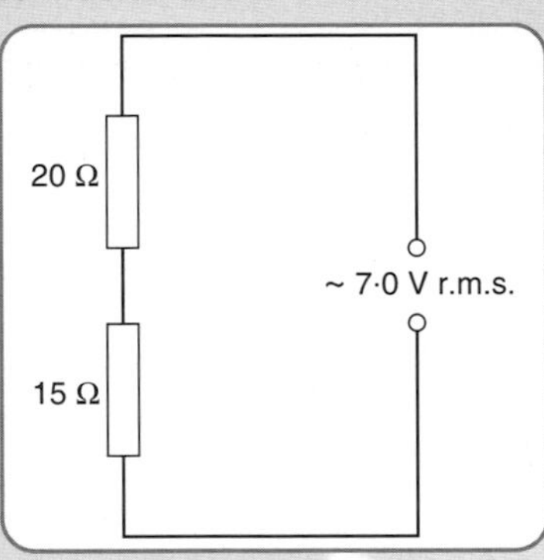

(a) Calculate the r.m.s. current in the 20 Ω resistor.

(b) Calculate the peak current in the 15 Ω resistor.

(c) Calculate the power dissipated in the 15 Ω resistor.

3 The a.c supply in the circuit below has an internal resistance of 2·2 Ω.

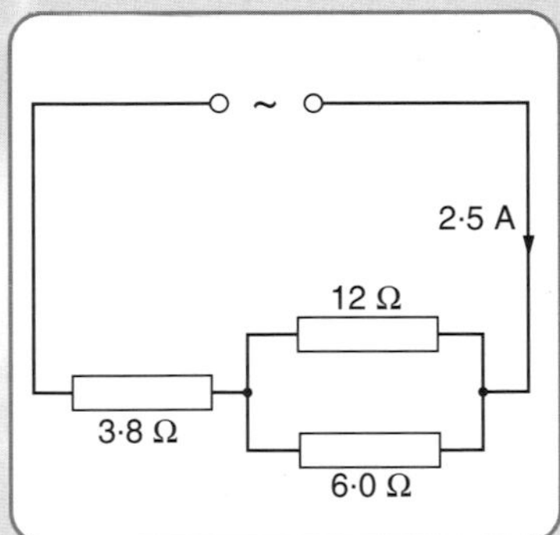

The r.m.s. current in the 3·8 Ω resistor is 2·5 A. Calculate the peak value of the e.m.f. of the supply.

***Exercises** continued* ➢

Exercises continued

4 The output of a signal generator is connected to the input terminals of an oscilloscope. The output frequency of the signal generator is set at 200 Hz. The trace obtained on the oscilloscope screen is shown below.

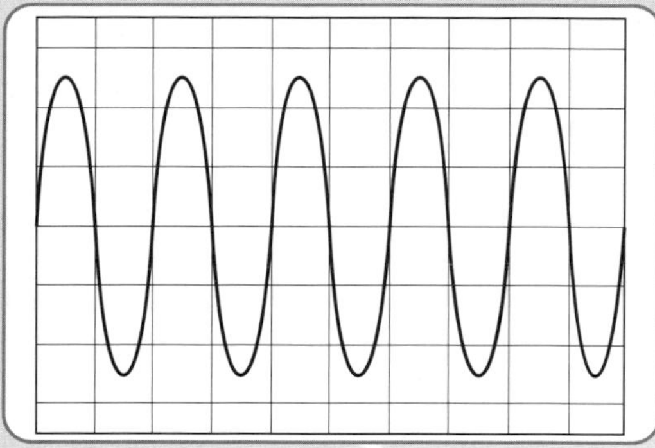

The Y-gain setting is 2 V div^{-1}.

(a) Calculate the peak voltage of the a.c. signal.

(b) Calculate the time-base setting of the oscilloscope in ms div^{-1}.

(c) With the oscilloscope settings unaltered, the output frequency of the signal generator is changed to 500 Hz. Describe the effect(s) that this has on the trace on the oscilloscope screen.

2.6 Capacitance

What You Should Know

For Higher Physics you need to be able to:

- understand and use the quantity capacitance
- explain why work must be done to charge a capacitor
- understand and use the relationships $C = \frac{Q}{V}$ and $E = ½QV$.

When two parallel conducting plates are charged there is an electric field between the plates and a p.d. across the plates. The excess charge on each plate is directly proportional to the p.d. across the plates. 'Excess charge' is usually shortened to 'charge' in Higher Physics papers.

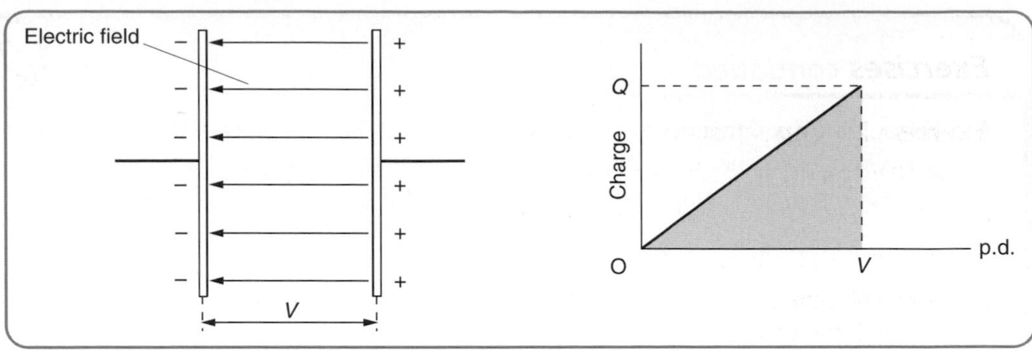

Figure 2.10

An arrangement of two parallel plates like the one shown in Figure 2.10 is called a capacitor.

Capacitance is the ratio of charge to p.d.: $C = \frac{Q}{V}$.

The unit of capacitance is the farad. 1 farad = 1 coulomb per volt = 1 C V^{-1}.

Capacitors and energy

Consider two initially uncharged parallel plates A and B. Negative charge is moved from A to B. A becomes positively charged and B becomes negatively charged. The negative charge is attracted to A and repelled from B. Work must be done to overcome the electrostatic forces between the plates and the charge.

The work done in charging a capacitor is equal to the area under the graph of charge against p.d. (see Figure 2.10).

W = area of shaded triangle $\Rightarrow W = \frac{1}{2}QV$.

The work done in charging a capacitor is stored as potential energy in the capacitor

$\Rightarrow$ energy stored in a capacitor $E = \frac{1}{2}QV$.

$C = \frac{Q}{V} \Rightarrow Q = VC$.

Substituting for Q in the above equation $\Rightarrow E = \frac{1}{2}VC \times V = \frac{1}{2}CV^2$.

Also $V = \frac{Q}{C}$.

Substituting for V in the above equation $\Rightarrow E = \frac{1}{2}Q \times \frac{Q}{C} = \frac{1}{2}\,\frac{Q^2}{C}$.

! Learn these relationships. Choosing the right version can save you time during your Higher Physics exam.

Exercises

Exercise 29 Capacitance

1 A 40 μF capacitor is connected to a 20 V d.c. supply. Calculate the
   (a) maximum charge stored in the capacitor
   (b) maximum energy stored in the capacitor.

2 A 250 nF capacitor is connected to a 120 V d.c. supply. Calculate the maximum energy stored in the capacitor.

3 'The excess charge on each plate is directly proportional to the p.d. across the plates.' Why is the word 'excess' included in this statement?

4 An initially uncharged 0·50 mF capacitor is charged by a steady current of 200 μA for 20 s. Calculate the final p.d. across the capacitor.

Experiments

Showing that $Q \propto V$

In your Higher Physics you could be asked about an experiment to show that $Q \propto V$. One method is described below.

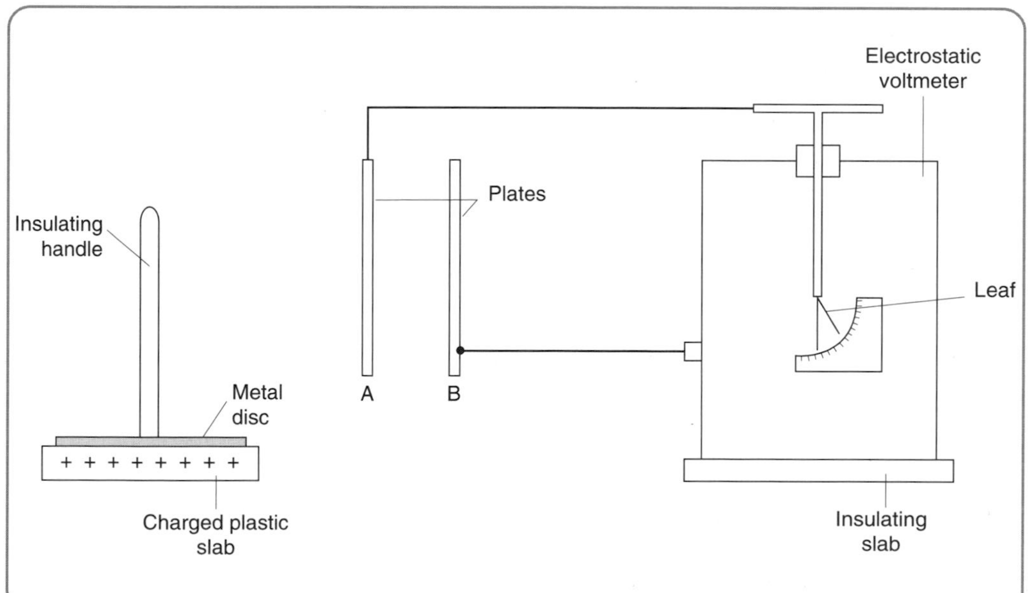

Figure 2.11

The plates are initially uncharged and the deflection of the leaf in the electrostatic voltmeter (electroscope) is zero.

***Experiments** continued* ➢

Experiments continued

The metal disc is earthed momentarily. This makes the disc negatively charged. The disc is lifted by the insulating handle and touched against plate A. Plate A becomes negatively charged and plate B becomes positively charged.

The deflection of the leaf is noted.

The process is repeated. Each time the metal disc is touched against plate A an approximately equal quantity of negative charge is transferred onto the plate.

The number of times charge is transferred and the deflection of the leaf are noted.

A graph of the number of times charge is transferred (charge) against the deflection of the leaf (p.d.) is plotted. A straight line graph passing through the origin is obtained. This shows that the p.d. across the plates of a capacitor is directly proportional to the excess charge on the plates.

2.7 Capacitors in circuits

What You Should Know

For Higher Physics you need to be able to:

- draw current–time and voltage–time graphs for the charging and discharging of capacitors in d.c. circuits
- state that in an a.c. circuit containing only capacitors current is directly proportional to the frequency of the supply
- solve problems on capacitors in circuits
- describe and explain possible uses of capacitors.

Capacitors in d.c. circuits

An initially uncharged capacitor is connected in the circuit shown in Figure 2.12.

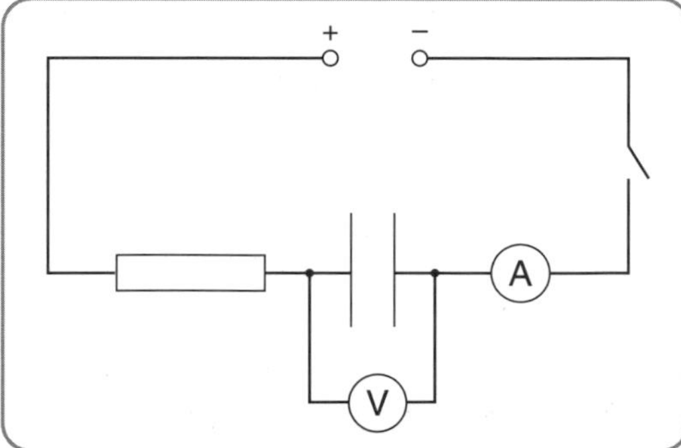

Figure 2.12

When the switch is closed, the ammeter reading starts high and gradually falls to zero. The voltmeter reading starts at zero and gradually increases to the e.m.f. of the supply. The graphs in Figure 2.13 show the current in the capacitor and the p.d. across the capacitor during charging.

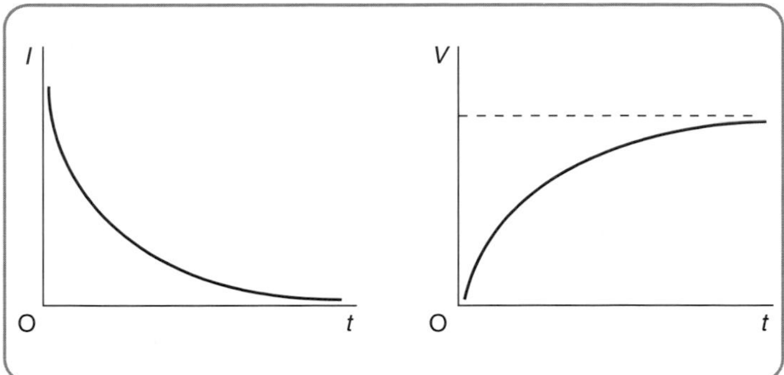

Figure 2.13

The p.d. across the capacitor acts in the opposite direction to the e.m.f. of the battery.

For your Higher Physics you need to be able to draw the graphs for charging and discharging capacitors. You could also be asked questions about capacitors that are at the beginning or part way through the charging or discharging process.

Capacitors in a.c. circuits

An initially uncharged capacitor is connected in the circuit shown in Figure 2.14.

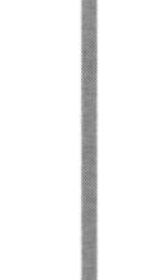

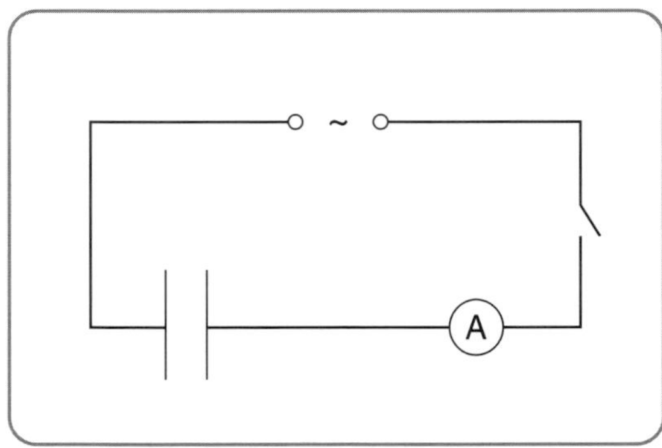

Figure 2.14

When the switch is closed, a steady reading is observed on the a.c. ammeter.

The frequency of the a.c supply is increased. The reading on the a.c. ammeter increases.

In a circuit containing only capacitors, the current is directly proportional to the frequency of the a.c. supply. The circuit above can be used to demonstrate this relationship.

Uses of capacitors

In Higher Physics you could be asked to describe and/or explain the following uses of capacitors.

Storing energy: When a capacitor is charged, energy is stored in the electric field between its plates. This energy can be used to provide a short high energy pulse of current. A good example is the flash unit of a camera. A capacitor is charged from the camera battery. When the shutter is pressed the capacitor discharges through the bulb giving an intense pulse of light.

Storing charge: A capacitor connected across the output of a d.c. power supply smooths the output, i.e. it reduces the size of any ripples in the output. The p.d. across the capacitor stabilises the output of the supply because of the time needed to move charge onto and off the plates.

Blocking d.c. while passing a.c.: It is possible to remove the d.c. part of a mixed signal that is part a.c. and part d.c. by passing the signal through a capacitor in series. A capacitor in series with a d.c. supply charges up to the e.m.f. of the supply. The p.d. across the capacitor acts in the opposite direction to the e.m.f. and so blocks the e.m.f. The a.c. part of the signal is allowed to pass.

Exercises

Exercise 30 Capacitors in circuits

1 An initially charged capacitor is connected in the circuit below.

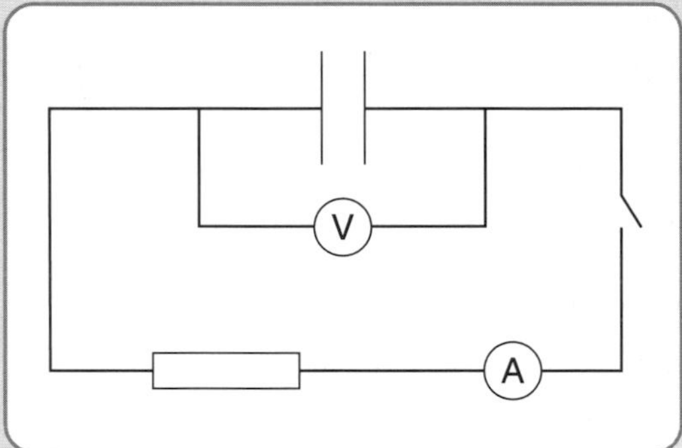

The switch is closed.

(a) Describe what happens to the readings on the meters.

(b) Sketch graphs to show the

(i) current in the capacitor during discharging (values are not required)

(ii) p.d. across the capacitor during discharging (values are not required).

2 An initially uncharged capacitor is connected in the circuit shown below.

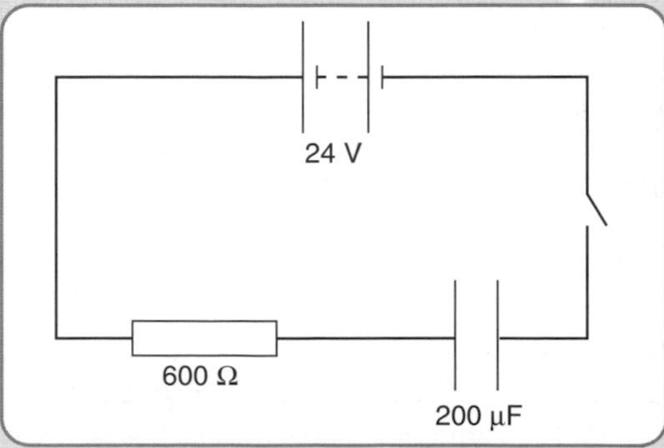

Exercises *continued* ➢

Exercises continued

The internal resistance of the battery is negligible. The switch is closed.

(a) State the initial p.d. across the capacitor.

(b) Calculate the initial current in the circuit.

(c) Calculate the p.d. across the capacitor when the current is 0.010 A.

(d) State the final p.d. across the resistor.

(e) State the final p.d. across the capacitor.

(f) Calculate the energy stored in the capacitor when it is fully charged.

3 Sketch a graph to show the relationship between current and frequency in an a.c. circuit containing only capacitors.

2.8 Analogue electronics: op-amps in inverting mode

What You Should Know

For Higher Physics you need to be able to:

- describe characteristics of an ideal op-amp
- identify circuits where an op-amp is being used in inverting mode
- solve problems on op-amps being used in inverting mode.

Operational amplifiers (op-amp)

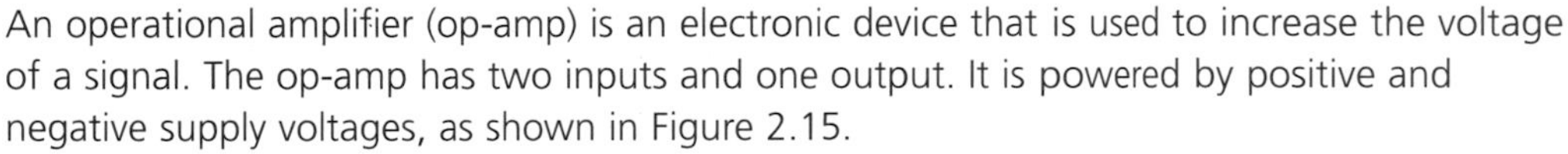

An operational amplifier (op-amp) is an electronic device that is used to increase the voltage of a signal. The op-amp has two inputs and one output. It is powered by positive and negative supply voltages, as shown in Figure 2.15.

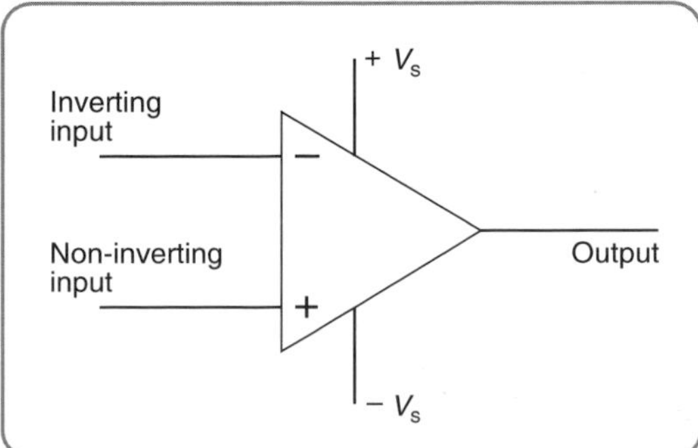

Figure 2.15

The power supply connections are often omitted from op-amp circuit diagrams in Higher Physics papers.

For an ideal op-amp:

- The input current is zero – this is equivalent to the op-amp having an infinite input resistance.
- Both inputs are at the same potential.

The behaviour of real op-amps is very close to ideal.

Op-amp in inverting mode

The circuit in Figure 2.16 shows an op-amp connected in inverting mode.

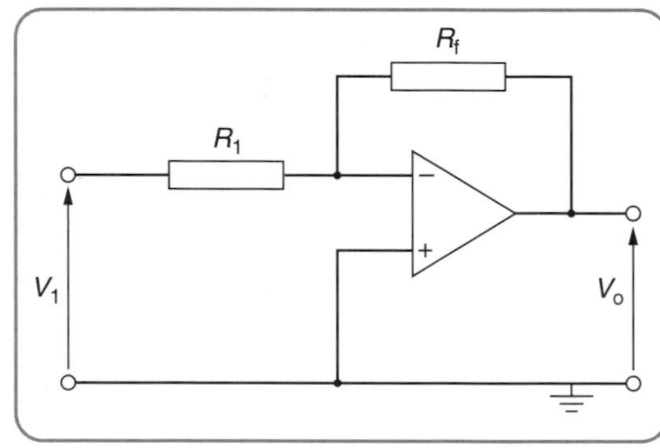

Figure 2.16

The relationship between the input voltage V_1 and output voltage V_o is $\frac{V_0}{V_1} = -\frac{R_f}{R_1}$.

The negative sign means that the input signal is inverted.

Values for the resistors R_f and R_1 are chosen to control the degree of amplification of the input signal. Input signals can be made bigger (amplified) or smaller (attenuated).

Saturation

An op-amp cannot produce an output voltage that is greater than the positive supply voltage or less than the negative supply voltage. When the output voltage reaches the supply voltage the amplifier is saturated.

Note: It is **incorrect** to say that the voltage is saturated – if you write this you will lose marks. It is the **op-amp** which is saturated.

In practice the amplifier becomes saturated when the size of the output voltage is a little less than the size of the supply voltage. For example, when the supply voltage is ±15 V the amplifier becomes saturated when the output is around ±13 V.

When an a.c. signal is applied to the inverting input and the op-amp is saturated, the output signal approximates to a square wave.

Exercises

Exercise 31 Op-amps in inverting mode

1 The operational amplifier connected in the circuit below is powered by a supply of ±15 V.

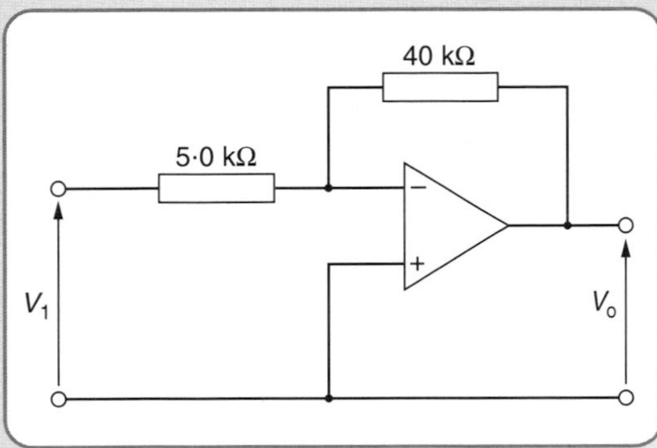

The output voltage V_o is – 12 V.

(a) Calculate the input voltage.

(b) The input voltage is now doubled. Calculate the new output voltage.

2 The supply voltage of the op-amp in the circuit below is ± 12 V.

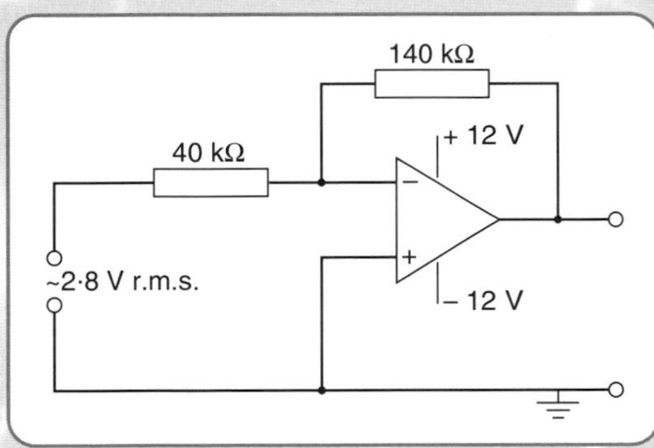

A 2·8 V r.m.s. a.c. supply is connected to the inverting input of the op-amp.

(a) Calculate the peak output voltage.

(b) Draw a sketch showing the shape of the output voltage (values not required).

(c) The supply voltage is increased to ± 16 V. Describe and explain any effect this change has on the output voltage.

2.9 Analogue electronics: op-amps in differential mode

What You Should Know

For Higher Physics you need to be able to:

- identify circuits where an op-amp is being used in differential mode
- solve problems on op-amps being used in differential mode
- describe how to use an op-amp with a Wheatstone bridge containing resistive sensors
- describe how to use an op-amp to control electrical devices via a transistor.

Figure 2.17 shows an op-amp connected in differential mode.

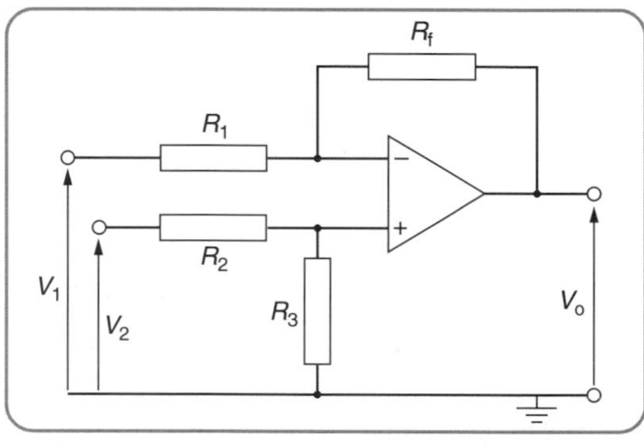

Figure 2.17

Resistor values are chosen so that $\frac{R_f}{R_1} = \frac{R_3}{R_2}$.

The relationship between input voltages V_1 and V_2 and output voltage is $V_o = (V_2 - V_1)\frac{R_f}{R_1}$.

In differential mode the op-amp amplifies the potential difference between its two inputs.

Values for the resistors R_f, R_1, R_2 and R_3 are chosen to control the degree of amplification of the potential difference. Potential differences between the inputs can be made bigger (amplified) or smaller (attenuated).

Saturation

An op-amp may also become saturated when it is connected in differential mode. Remember, the output cannot be greater than the positive supply voltage or less than the negative supply voltage.

Op-amps and the Wheatstone bridge

A common type of question in Higher Physics has an op-amp connected to a Wheatstone bridge circuit. One of the resistors in the Wheatstone bridge is usually a resistive sensor such as an LDR or a thermistor.

The resistive sensor could be at any of the four resistor locations in the Wheatstone bridge.

The circuits are complicated so handle the Wheatstone bridge and op-amp calculations separately. The result of one gives information that you use in the other. This is illustrated in the following example.

Questions and Answers

An op-amp is connected to a Wheatstone bridge circuit as shown below.

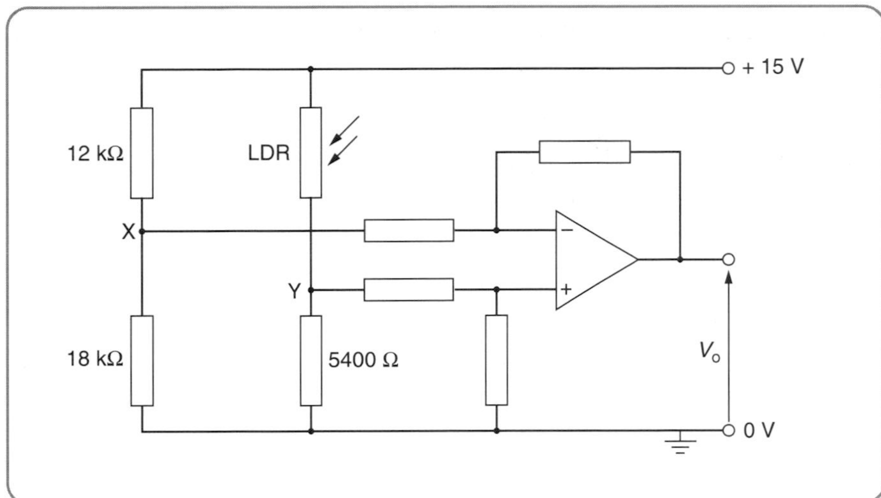

(a) Calculate the resistance of the LDR when the output voltage of the op-amp is zero.

(b) Explain whether the output voltage is positive or negative when the resistance of the LDR is 4000 Ω.

(c) Describe and explain how the output voltage changes when the resistance of the LDR changes from 4000 Ω to 3000 Ω.

(a) The op-amp is in differential mode. It is amplifying the p.d. between X and Y.

V_o is zero ⇒ the p.d. between X and Y is zero ⇒ the Wheatstone bridge is balanced.

$R_1 = 12\ \text{k}\Omega = 12\,000\ \Omega \qquad \frac{R_1}{R_2} = \frac{R_3}{R_4}$

$R_2 = 18\ \text{k}\Omega = 18\,000\ \Omega \quad \Rightarrow \quad \frac{12\,000}{18\,000} = \frac{R_3}{5400}$

$R_4 = 5400\ \Omega \quad \Rightarrow \quad R_3 = 3600\ \Omega$

Note: Compare this with the Wheatstone bridge in question 2 of *Exercise 27 Wheatstone bridge*.

(b) When the resistance of the LDR is 3600 Ω the voltage at Y = voltage at X.
When the resistance of the LDR is 4000 Ω the voltage at Y < voltage at X.

$V_o = (V_2 - V_1)\frac{R_f}{R_1}$. $V_2 < V_1 \Rightarrow (V_2 - V_1)$ is negative $\Rightarrow V_o$ is negative.

(c) The output voltage changes (increases) from negative to positive when the resistance of the LDR is changed from 4000 Ω to 3000 Ω.

The voltage at X is constant. As the resistance of the LDR is decreased the voltage at Y increases from a value less than the voltage at X to a value greater than the voltage at X.

Automatic control of electronic circuits

The change in the output voltage of an op-amp from negative to positive can be used to switch a transistor on. Similarly a change from positive to negative can be used to switch a transistor off. These changes are used in circuits that control electrical devices.

This type of circuit has three distinct parts: the Wheatstone bridge, the op-amp and the transistor. The example below illustrates how you go about answering a question like this.

Questions and Answers

The circuit below is used to switch on the heating in a greenhouse when the temperature in the greenhouse falls below 24 °C.

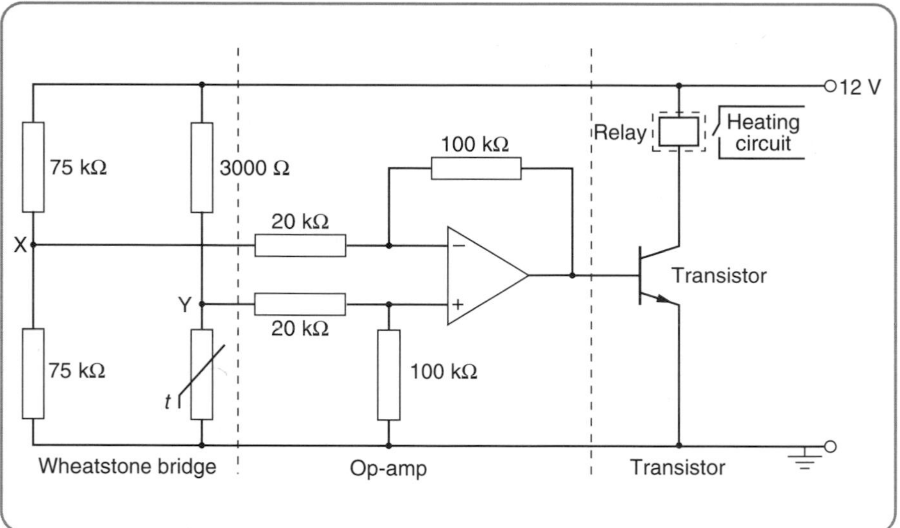

The initial temperature in the greenhouse is 26 °C. The resistance of the thermistor is 2900 Ω at 26 °C.

(a) Calculate the output voltage of the op-amp at 26 °C. State any assumption you make in your calculation.

(b) The transistor switches on when the output voltage of the op-amp is 0·70 V. Calculate the resistance of the thermistor when the transistor switches on.

(c) Describe and explain how the circuit operates to switch on the heating.

(d) Describe how the circuit operates after the heating is switched on.

Questions *and* ***Answers*** *continued* ➢

Questions and Answers continued

(a) p.d. across the thermistor $= V \times \dfrac{R_1}{R_1 + R_2} = 12 \times \dfrac{2900}{2900 + 3000} = 5{\cdot}898 \text{ V} = 5{\cdot}9 \text{ V}.$

p.d. across the lower 75 kΩ resistor $= 12 \times \dfrac{75\,000}{75\,000 + 75\,000} = 6{\cdot}0 \text{ V}.$

$$V_o = (V_2 - V_1)\frac{R_f}{R_1}$$

$$= (5{\cdot}9 - 6{\cdot}0) \times \frac{100 \times 10^3}{20 \times 10^3}$$ *p.d.s must be substituted correctly*

$$= -\,0{\cdot}5 \text{ V}$$ *note the negative sign!*

Assumption: The currents between X and Y and the op-amp part of the circuit are negligible (*remember that for an ideal op-amp the input current is zero*).

(b) $V_o = 0{\cdot}70$ V

$$V_o = (V_2 - V_1)\frac{R_f}{R_1}$$

$$\Rightarrow \quad 0{\cdot}70 = (V_2 - 6{\cdot}0) \times \frac{100 \times 10^3}{20 \times 10^3}$$

$$\Rightarrow \quad V_2 = 6{\cdot}14 \text{ V}$$

p.d. across the thermistor $= V \times \dfrac{R_1}{R_1 + R_2}$

$$\Rightarrow \quad 6{\cdot}14 = 12 \times \left(\frac{R_1}{R_1 + 3000}\right)$$

$$\Rightarrow \quad R_1 = 3143 = 3100\ \Omega$$

(c) At 26 °C the output voltage of the op-amp is negative and the transistor is off. As the temperature falls the resistance of the thermistor increases. The output voltage of the op-amp increases.

When the resistance of the thermistor is 3100 Ω the output voltage of the op-amp is +0·70 V and the transistor is switched on.

This switches on the relay and the relay switches on the heating circuit.

(d) When the heating is on the temperature increases. The resistance of the thermistor decreases and the output voltage of the op-amp decreases. When the output voltage of the op-amp falls below +0·70 V the transistor is switched off. This switches off the relay and the heating is switched off.

Exercises

Exercise 32 Op-amps in differential mode

1 The circuit shown is designed for an alarm system.

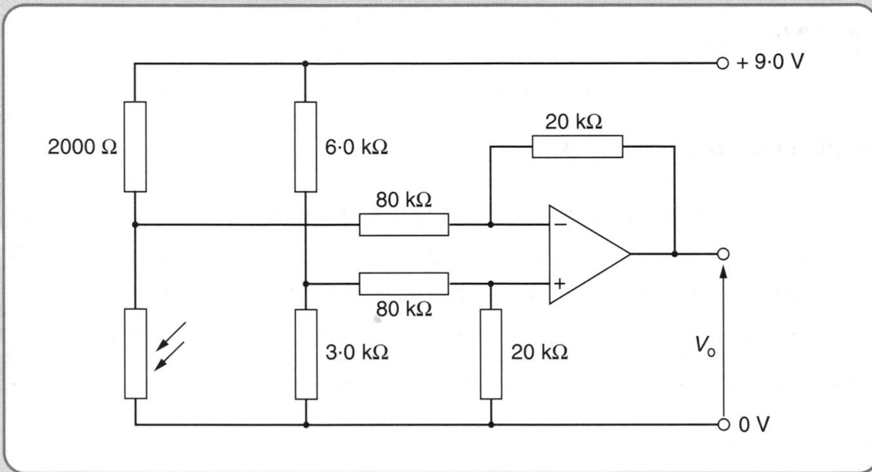

The LDR is illuminated by dim light and $V_o = -0{\cdot}75$ V.

(a) Calculate the p.d. across the LDR when $V_o = -0{\cdot}75$ V.

(b) Calculate the resistance of the LDR when $V_o = -0{\cdot}75$ V.

(c) The intensity of light incident on the LDR is gradually increased until the LDR is illuminated by bright light. Describe and explain the effect this has on the output voltage of the op-amp.

2 The circuit below is designed to switch on a heating circuit when the temperature falls below a certain value.

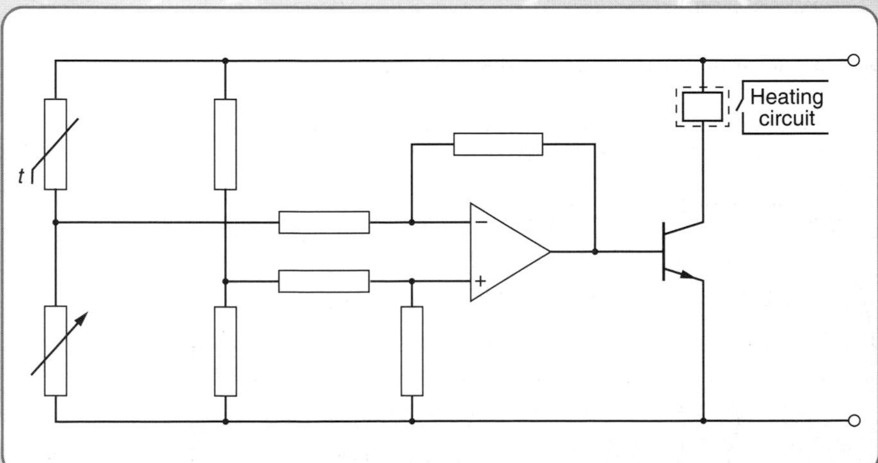

(a) Explain how the circuit operates to switch on the heating circuit.

(b) State the function of the variable resistor in this circuit. Explain your answer.

Chapter 3

RADIATION

3.1 Waves

What You Should Know

For Higher Physics you need to be able to:

- understand and use the quantities frequency, period, wavelength and amplitude
- understand and use the terms in phase, out of phase, and coherent
- understand the characteristic behaviours of waves: reflection, refraction and diffraction.

Wave properties

A wave is a regular vibration that carries energy through a medium.

The medium is the 'material' the wave is passing through. Usually the medium is a material like air or water or iron. Sometimes in Higher Physics questions the medium for electromagnetic waves is a vacuum.

Frequency, f, is the number of waves per unit time. This is the same as the frequency of the source of the waves.

Period, T, is the time taken to produce one wave: $T = \frac{1}{f}$.

You should also remember the wave equation $v = f\lambda$.

Figure 3.1 illustrates wavelength and amplitude.

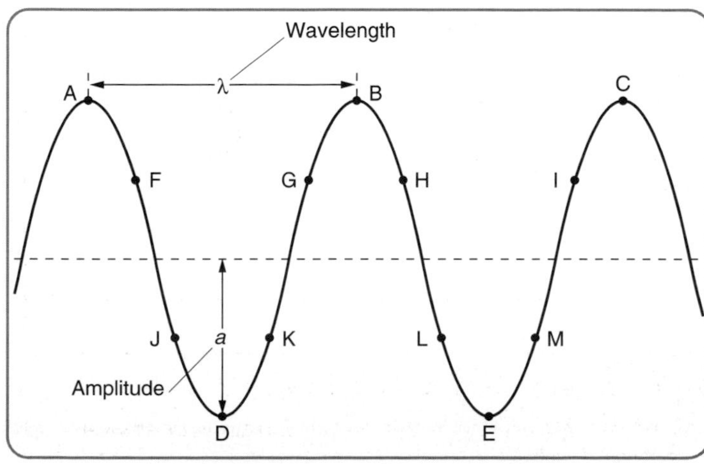

Figure 3.1

Amplitude, a, is the maximum displacement of the medium from its rest position. (The rest position is the position of the medium when there is no wave present.) The greater the amplitude the greater the energy carried by the wave.

Wavelength, λ, is the length of one complete wave measured parallel to the direction in which the wave is travelling.

Any two points a whole number of wavelengths apart are in phase. For example, in the diagram above points A, B and C are in phase.

Any two points an odd number of half wavelengths apart are out of phase. For example, in the diagram above points B and E are out of phase.

Two sources of waves are **coherent** when

- they have the same frequency
- there is a constant phase difference between them.

Wave behaviour

Reflection, refraction and diffraction are characteristic behaviours of all types of wave. **Reflection** occurs when waves bounce from the surface of an object (see Figure 3.2).

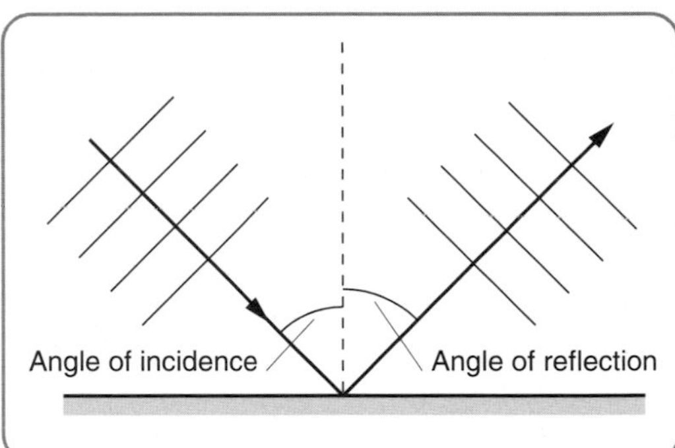

Figure 3.2

None of the wave properties frequency, period, wavelength or speed change during reflection. The only change is the direction of the wave.

When a wave is reflected

$$\text{angle of incidence} = \text{angle of reflection}.$$

For objects with irregular shapes the waves are reflected in all directions. Reflected light enables us to see objects that do not emit light.

Refraction occurs when a wave moves from one medium to another (see Figure 3.3).

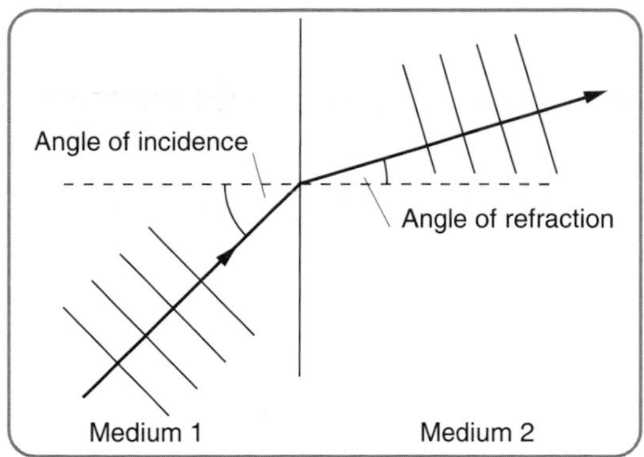

Figure 3.3

The frequency and period of a wave do not change during refraction.

Wave speed is determined by the medium and so the speed of a wave usually changes during refraction. The change in speed causes the wavelength to change. When the wave gets faster the wavelength gets bigger; when the wave gets slower the wavelength gets smaller.

When a wave is incident at any angle other than normal, the direction of the wave also changes.

Diffraction occurs when waves pass an edge, pass an object or go through a gap (see Figure 3.4).

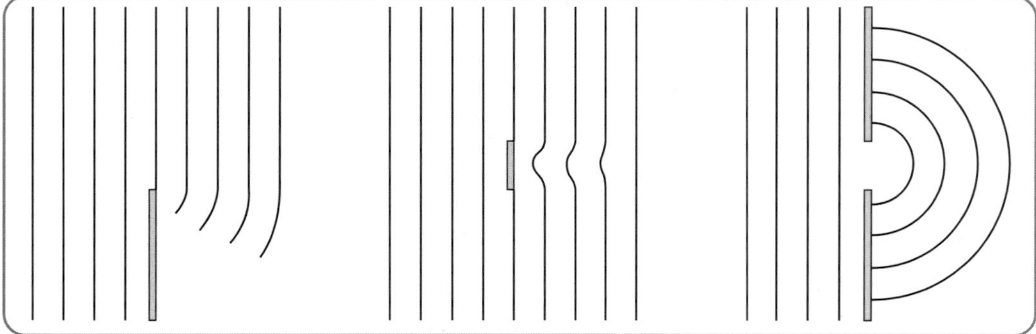

Figure 3.4

None of the wave properties frequency, period, wavelength or speed change during diffraction. The only change is the direction of the wave.

When a wave passes through a gap or goes past an object, the diffraction is greatest when the width of the gap or object is about the same size as or less than the wavelength of the wave.

At an edge, long wavelength waves diffract more than short wavelength waves.

Exercises

Exercise 33 Waves

1 A girl makes waves on the surface of a pool by tapping the surface with a frequency of 2·0 Hz.

(a) State the frequency of the waves.

(b) Calculate the period of the waves.

(c) A boy is standing on the edge of the pool. How many waves pass the boy each second?

(d) The wavelength of the waves is 30 cm. Calculate the speed of the waves.

(e) The girl increases the frequency to 3·0 Hz. Calculate the new wavelength of the waves.

2 (a) In Figure 3.1, the wavelength of the wave is shown between points A and B. Identify four other pairs of points that are one wavelength apart.

(b) Points B and C are in phase. Name four other pairs of points that are in phase.

(c) Points A and D are out of phase. Name four other pairs of points that are out of phase.

3 Copy and complete the following table to show which wave properties change when waves are reflected, refracted and diffracted.

Behaviour	*Frequency*	*Period*	*Wavelength*	*Speed*
Reflection	No change			
Refraction				Changes
Diffraction			No change	

4 Explain the following in terms of wave behaviour.

(a) Your friend hides behind a tree and calls your name – you cannot see your friend but you can hear your name being called.

(b) A straight piece of wood looks as if it bends when it is partially immersed in clear pond water.

(c) You can see this book and these words.

(d) The colour of a dress looks different when inside a shop and in daylight outside the shop.

5 Radio waves of frequency 102·5 MHz are reflected by a satellite dish.

(a) Calculate the period of these waves.

(b) Calculate the wavelength of these waves in a vacuum.

(c) State the frequency of the radio waves after they are reflected by the dish.

3.2 Interference

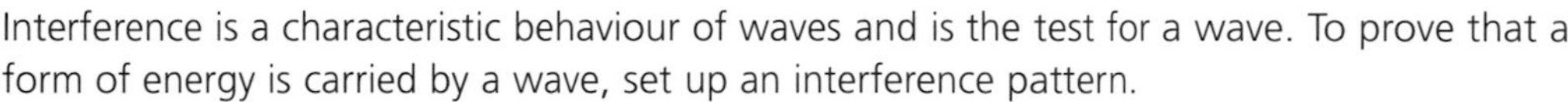

What You Should Know

For Higher Physics you need to be able to:

- state that interference is the test for a wave
- explain interference in terms of crests and troughs
- state the conditions for the maxima and minima of an interference pattern.

Interference

Interference is a characteristic behaviour of waves and is the test for a wave. To prove that a form of energy is carried by a wave, set up an interference pattern.

Two coherent sources of waves are needed to produce an interference pattern. Remember that coherent sources have the same frequency and there is a constant phase difference between them.

Interference occurs when waves from two sources overlap in the same region of space – this is called superposition of waves.

Constructive interference occurs when the waves are in phase. When a crest from one source meets a crest from the other source, a bigger crest is formed.

Destructive interference occurs when the waves are out of phase. When a crest from one source meets a trough from the other source a smaller wave is produced. When the amplitudes of the crest and trough are equal they cancel out to produce a point where there is no disturbance.

When the sources are coherent, the points of constructive interference and destructive interference form a pattern.

Experiments

Double-slit experiments

The Higher Physics examination frequently includes questions on double-slit experiments (a double-slit is two slits close together). In these experiments the double-slit forms two coherent sources of waves. The experiment is set up as shown in Figure 3.5.

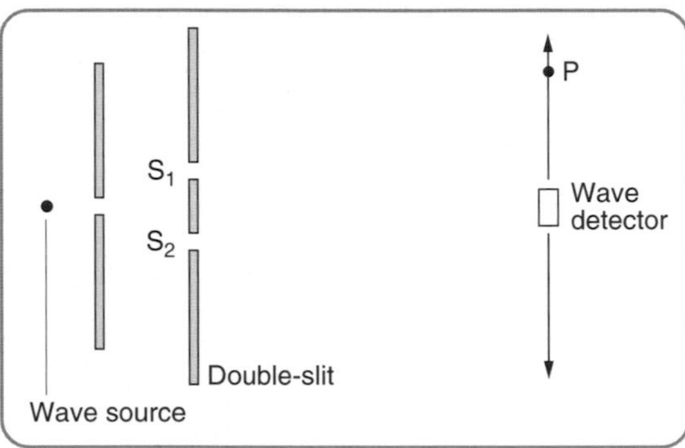

Figure 3.5

Waves are first passed through a single slit and then a double-slit.

Slits S_1 and S_2 are coherent sources. The phase difference between S_1 and S_2 is assumed to be negligible.

A wave detector is moved across the space to the right of the double-slit to find the pattern of constructive and destructive interference. For visible light a screen is used to show the interference pattern.

For constructive interference the path difference ($S_1P - S_2P$) is a whole number of wavelengths.

$$\text{path difference} = n\lambda, \text{ where } n \text{ is an integer}$$

For destructive interference the path difference is an odd number of half wavelengths.

$$\text{path difference} = (n + \tfrac{1}{2})\lambda, \text{ where } n \text{ is an integer}$$

Exercises

Exercise 34 Interference

1 Microwave radiation of wavelength 4·0 cm is incident on a metal plate which has two slits P and Q as shown below.

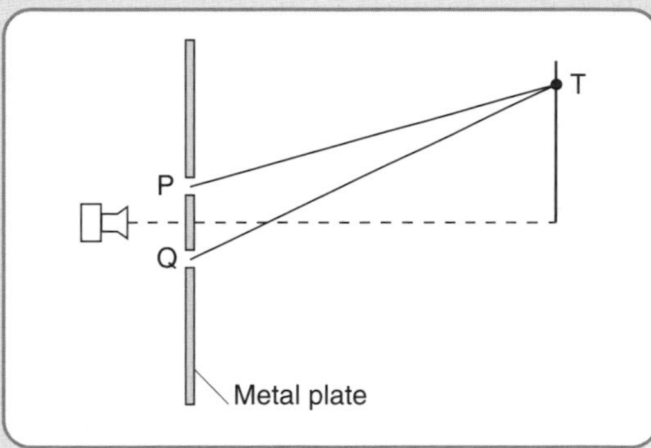

(a) What type or types of interference occur on the broken line in the diagram? Explain your answer.

(b) PT = 0·640 m and QT = 0·760 m. What type of interference occurs at point T? Justify your answer by calculation.

(c) How many points of destructive interference are there on the straight line between point T and the broken line in the diagram? Explain your answer.

2 Monochromatic light of wavelength 600 nm is incident on a double-slit.

(a) State three values of path difference that produce constructive interference.

(b) State three values of path difference that produce destructive interference.

3.3 Gratings and spectra

What You Should Know

For Higher Physics you need to be able to:

- describe the effect of a grating on a monochromatic light beam
- describe how to measure the wavelength of light using a grating
- state approximate values for the wavelengths of red, green and blue light
- describe and compare white light spectra produced by a grating and a prism.

Gratings

A grating consists of many slits (lines) close together and a constant distance apart.

When waves are incident on a grating all of the lines act as coherent sources of the waves. Compared with a double slit, the interference pattern produced by a grating has fewer, more widely spaced points of constructive interference. Figure 3.6 shows the effect of a grating on monochromatic light.

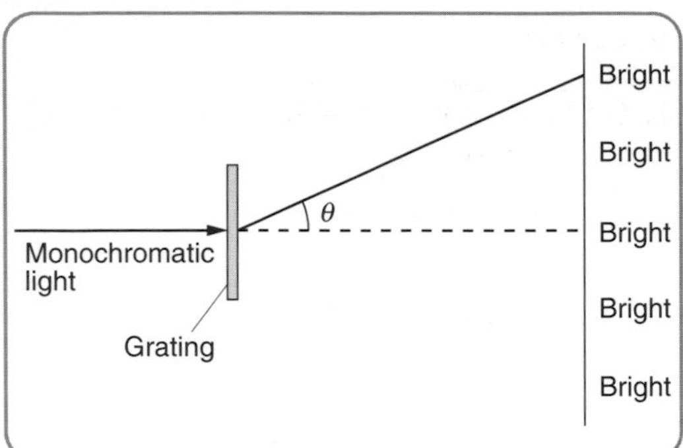

Figure 3.6

The bright band in the centre is called the zero order maximum. The order numbers of the other maxima are obtained by counting from the zero order.

For maxima, the relationship between the wavelength of light λ and angle θ is

$$d \sin \theta = n\lambda,$$

where d is the spacing of grating lines and n is the order number of the maximum. This is the grating equation. For the example in Figure 3.6, $d \sin \theta = 2\lambda$.

Experiments

Measuring the wavelength of light

To measure the wavelength of monochromatic light, a parallel beam of the light is fired at a grating mounted on the turntable of a spectrometer, as shown in Figure 3.7.

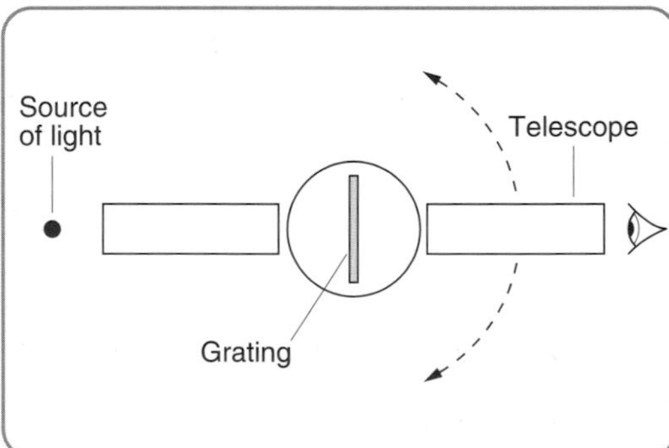

Figure 3.7

The position of the zero order maximum is noted. The telescope is then rotated one way. The order number and angle at the centre of each bright line are noted. The telescope is rotated the other way and the measurements are repeated. The average angle is calculated for each order.

The grating equation is used to calculate a value of λ for each order. A mean value of the wavelength is then calculated.

Ranges of values for the wavelengths of red, green and blue light are given in Table 3.1. For Higher Physics you need to be able to state wavelengths for each of these colours.

Table 3.1 Wavelengths for colours

Colour	*Wavelength/nm*
Red	620 to 700
Green	490 to 580
Blue	450 to 490

White light spectra

When white light is incident on a grating the zero order maximum is white. Visible spectra are produced at the other orders.

From the grating relationship $\sin\theta = \dfrac{n\lambda}{d}$

d is constant and for each order spectrum n is the same

$\Rightarrow \sin\theta$ is directly proportional to λ

$\Rightarrow$ the bigger the wavelength the bigger the angle θ.

Violet light has the shortest wavelength and is deviated least. Red light has the longest wavelength and is deviated most.

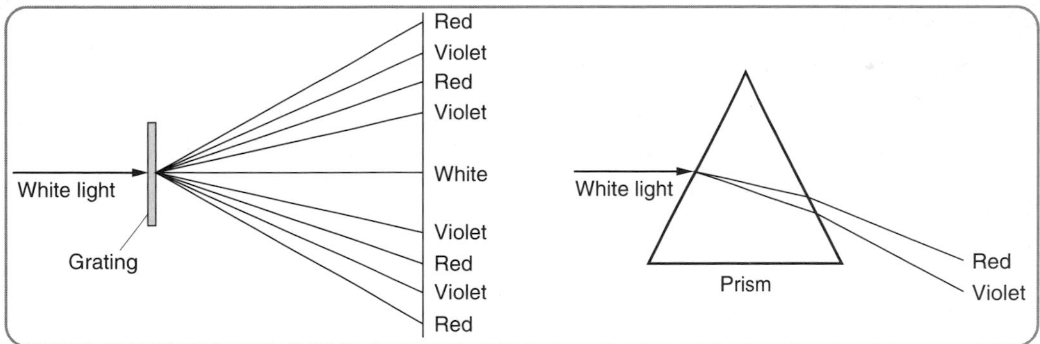

Figure 3.8

When a narrow beam of white light passes through a prism a visible spectrum is produced.

In this spectrum red light is deviated least and violet light is deviated most. This is the opposite to the deviation of the colours by a grating.

With a prism only one spectrum is produced. With a grating there are spectra on both sides of the zero order maximum.

Exercises

Exercise 35 Gratings and spectra

1 A light source of unknown frequency is incident on a grating. The grating has $2{\cdot}80 \times 10^5$ lines per metre.

The third-order maximum of intensity is at an angle of 25° to the central maximum.

(a) Calculate the wavelength of the light.

(b) Calculate the frequency of the light.

(c) State the colour of light of this frequency.

2 Monochromatic light of frequency $5{\cdot}20 \times 10^{14}$ Hz is incident on a grating. The spacing between lines in the grating is $4{\cdot}50 \times 10^{-6}$ m. Calculate the angle between the second-order maxima.

3 White light is incident on a grating which has $3{\cdot}20 \times 10^5$ lines per metre. Calculate the difference in the deviation between the violet and red ends of the first-order spectrum.

3.4 Refraction of light

What You Should Know

For Higher Physics you need to be able to:

- define refractive index for light passing from one medium to another
- describe how to measure the absolute refractive index of glass for monochromatic light
- state that refractive index depends on frequency
- understand and use all of the relationships for refractive index
- solve problems on refraction and refractive index.

When light passes from one medium to another, it changes speed. This is called **refraction**.

The absolute **refractive index** *n* of a medium is the ratio of the speed of light c in a vacuum (or air as an approximation) to the speed of light v in the medium.

$$n = \frac{\text{speed of light in a vacuum}}{\text{speed of light in the medium}} = \frac{c}{v}$$

Light travels faster in a vacuum than in any other medium. All absolute refractive indices are bigger than 1. The bigger the absolute refractive index the more slowly light travels in the medium.

When light travels from a medium 1 to a medium 2 the refractive index *n* is given by

$$n = \frac{v_1}{v_2},$$

where v_1 is the speed of light in medium 1 and v_2 is the speed of light in medium 2.

The frequency of light does not change during refraction

$$\Rightarrow \qquad n = \frac{v_1}{v_2} = \frac{f\lambda_1}{f\lambda_2} = \frac{\lambda_1}{\lambda_2}$$

where λ_1 is the wavelength in medium 1 and λ_2 is the wavelength in medium 2.

A ray of light incident at right angles to the surface between two optical media travels in a straight line. A ray incident at any other angle changes direction. (See Figure 3.9.)

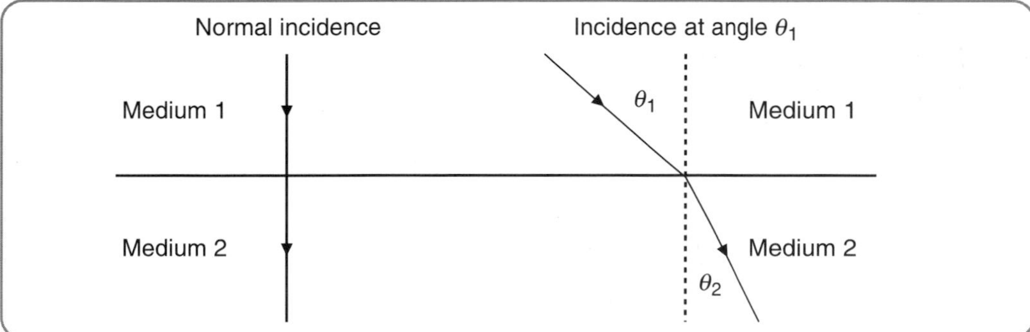

Figure 3.9

The ratio $\frac{\sin \theta_1}{\sin \theta_2}$ is constant and is equal to the refractive index: $n = \frac{\sin \theta_1}{\sin \theta_2}$.

When a ray bends towards the normal the light gets slower and the wavelength decreases.

When a ray bends away from the normal the light gets faster and the wavelength increases.

Experiments

Measuring absolute refractive index

The absolute refractive index of a medium for monochromatic light is measured using the apparatus shown in Figure 3.10.

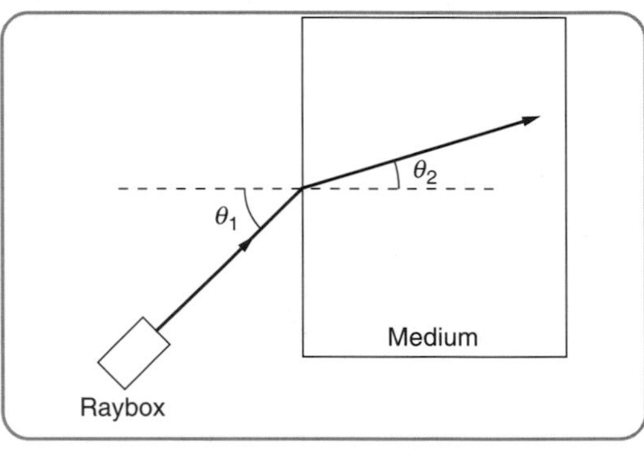

Figure 3.10

A narrow beam of the light is directed obliquely at the surface of a rectangular block of the medium. Angles θ_1 and θ_2 are measured. Angle θ_1 is altered several times and the measurements are repeated.

The ratio $\dfrac{\sin \theta_1}{\sin \theta_2}$ is calculated for each pair of angles. The average value is then calculated.

This is equal to the absolute refractive index of the medium for this light.

In this experiment angle θ_1 must always be the angle in air.

The absolute refractive index of a medium depends on the frequency of the light. For example, the refractive index of red light is less than the refractive index of violet light.

Exercises

Exercise 36 Refraction of light

1 Light of frequency $4{\cdot}58 \times 10^{14}$ Hz travels at a speed of $1{\cdot}24 \times 10^{8}$ m s^{-1} in diamond.

(a) Calculate the refractive index of diamond for this frequency of light.

(b) Calculate the wavelength of the light in diamond.

(c) State the frequency of this light in air.

***Exercises** continued* ➢

Exercises continued

2 A ray of light is incident on the surface of water as shown below.

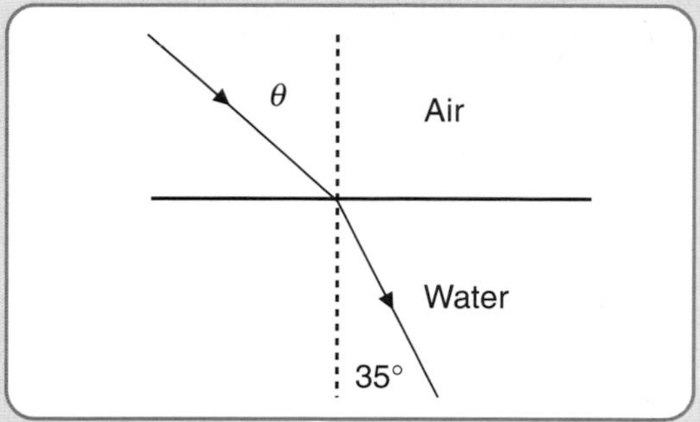

(a) State whether the light travels faster in air or water. Justify your answer.

(b) Calculate the angle of incidence.

3 A ray of light of frequency $4{\cdot}70 \times 10^{14}$ Hz is incident on the surface of a piece of clear plastic.

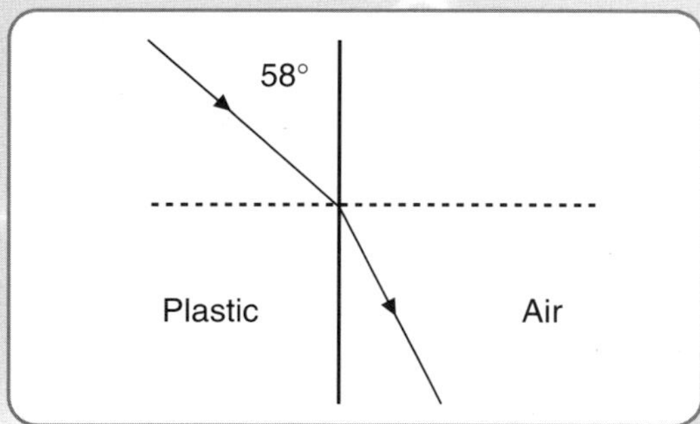

The refractive index of the plastic is 1·38 for this frequency of light.

(a) Calculate the angle of refraction.

(b) Calculate the wavelength of the light inside the plastic.

(c) State the colour of this light.

(d) Calculate the speed of the light inside the plastic.

4 A ray of white light is incident as shown below on one side of a triangular glass prism.

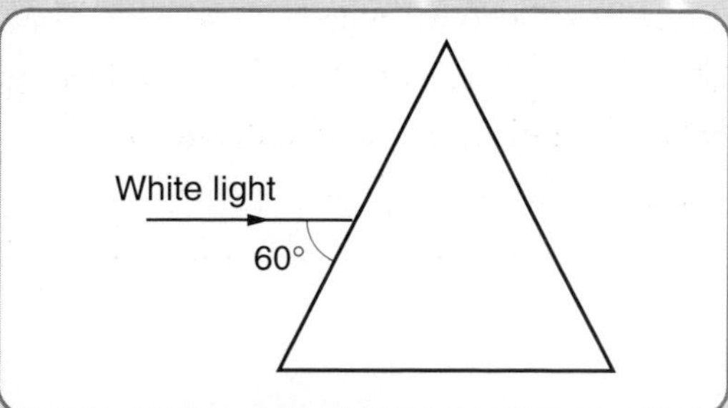

The refractive indices of the glass are 1·48 for red light and 1·52 for violet light. Calculate the angle between the red and violet ends of the spectrum inside the prism.

3.5 Total internal reflection and critical angle

What You Should Know

For Higher Physics you need to be able to:

- understand and use the terms total internal reflection and critical angle
- describe how to measure critical angle
- derive, understand and use the relationship between critical angle and refractive index
- solve problems on critical angle.

Figure 3.11 shows light incident on the glass side of a glass–air interface.

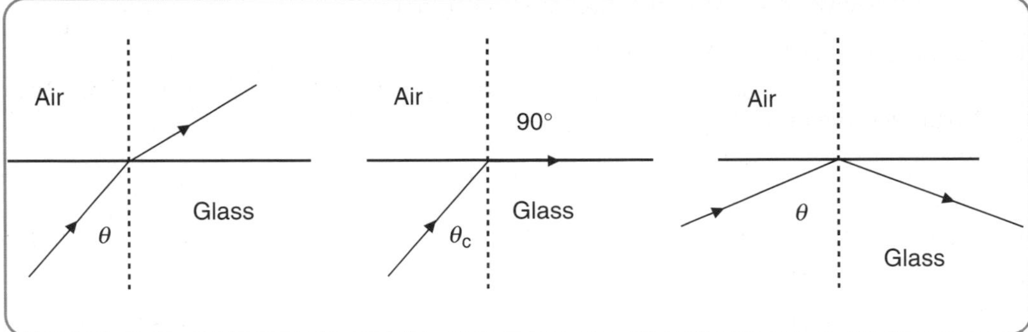

Figure 3.11

The light travels faster in air than it does in glass. The light gets faster when it is refracted. The angle of refraction is bigger than the angle of incidence θ.

The angle of incidence is increased until the angle of refraction is 90°. At this point, the angle of incidence is called the **critical angle**, θ_c.

When the angle of incidence is increased above the critical angle, the angle of refraction cannot increase any further. The light cannot pass through the surface and it is completely reflected.

This is called **total internal reflection**.

It is called total because all of the energy is reflected, internal because the energy stays inside the glass, and reflection because the light is reflected.

$\theta_1 = 90°, \theta_2 = \theta_c \qquad n = \dfrac{\sin \theta_1}{\sin \theta_2} = \dfrac{\sin 90°}{\sin \theta_c} = \dfrac{1}{\sin \theta_c}$

Experiments

Measuring critical angle

The critical angle of a medium for monochromatic light is measured using the apparatus shown in Figure 3.12.

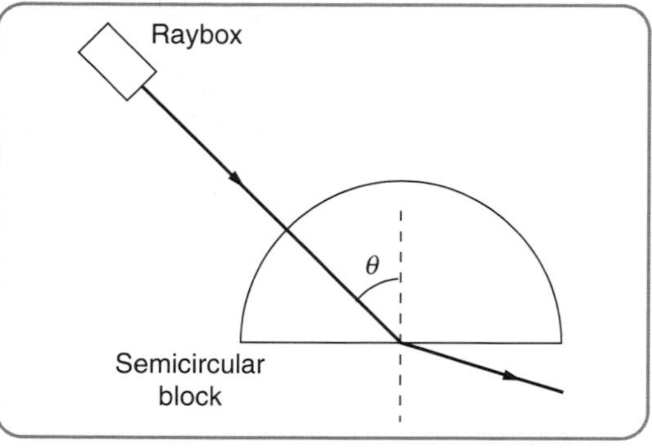

Figure 3.12

A narrow beam of the light is directed radially (i.e. along a radius) at the curved surface of a semicircular block of the medium.

The angle of incidence, θ, is gradually increased until total internal reflection occurs. The angle θ is measured. The angle of incidence is now gradually decreased until a refracted ray is observed.

The angle θ is again measured. The average of the two angles is the measured value of the critical angle.

The experiment is repeated several times and an average value of critical angle is calculated.

Exercises

Exercise 37 Total internal reflection and critical angle

1 Calculate the critical angle for red light incident on a water–air interface. The refractive index of water is 1·31 for this red light.

2 A ray of monochromatic light is incident on a glass–water interface. The refractive indices of the glass and water for this light are 1·46 and 1·30 respectively.

(a) Calculate the refractive index for the light travelling from water to glass.

(b) Calculate the critical angle for the refraction of the light at the glass–water interface.

3.6 Irradiance

What You Should Know

For Higher Physics you need to be able to:

- define irradiance
- describe how to show the relationship between irradiance and distance from a source
- solve problems on irradiance.

The irradiance *I* of radiation at a surface is the power per unit area incident on the surface.

$$I = \frac{P}{A}$$

The SI unit of irradiance is the watt per square metre, W m^{-2} (that is, J s^{-1} m^{-2}).

For a point source of radiation, irradiance is inversely proportional to the square of the distance from the point source, i.e.

$$I = \frac{k}{d^2} \Rightarrow Id^2 = k,$$ where *k* is a constant that depends on the source.

Experiments

Showing the relationship between irradiance and distance from a point source

The apparatus shown in Figure 3.13 is set up in a darkened laboratory.

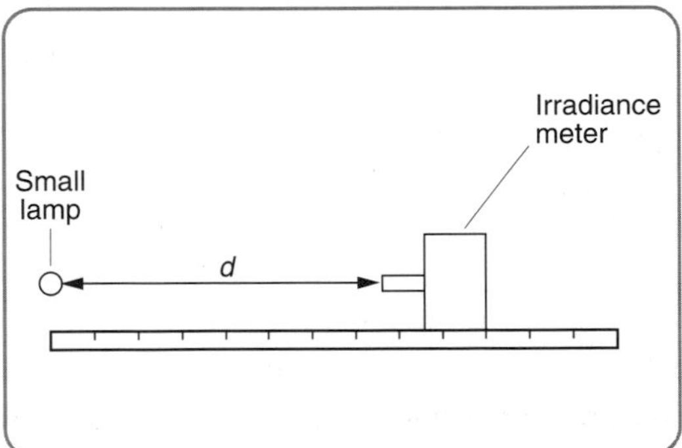

Figure 3.13

The distance *d* between the lamp and the irradiance meter is varied. Values of *d* and irradiance are noted.

A graph of I *against* $\frac{1}{d^2}$ is drawn. This is a straight line graph passing through the origin

$\Rightarrow I \propto \frac{1}{d^2}$.

Exercises

Exercise 38 Irradiance

1 The irradiance of light from a small lamp is 3·6 W m^{-2} at a distance of 2·0 m from the lamp. Calculate the irradiance at a distance of 6·0 m from the lamp.

2 A student sets up the apparatus below to investigate the irradiance of refracted rays.

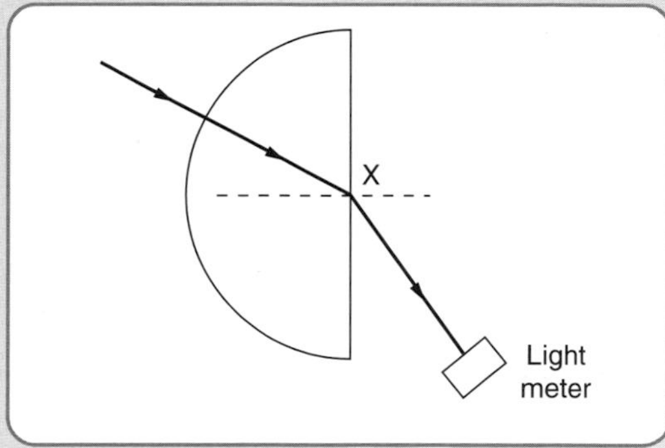

The angle of incidence is varied. The student uses the light meter to measure the irradiance due to the refracted ray at various angles of incidence. Explain why it is important for the student to keep the light meter at a constant distance from point X for each measurement of irradiance.

3.7 Photoelectric effect

What You Should Know

For Higher Physics you need to be able to:

- describe the conditions for the photoelectric effect
- describe the effects of increasing the irradiance of radiation at frequencies above and below the threshold frequency
- define irradiance in terms of photon energy
- understand and use the relationships between photon energy, frequency and the kinetic energy of photoelectrons
- solve problems on the photoelectric effect.

Electromagnetic radiation incident on the surface of a material may cause electrons to be ejected from the surface. This is called the photoelectric effect.

The photoelectric effect occurs *only* when the **frequency** of the incident radiation is **greater** than a **threshold** frequency f_o which depends on the nature of the surface.

Electrons ejected from the surface are called photoelectrons.

When the frequency of incident radiation is below the threshold frequency no photoelectrons are ejected no matter how great the irradiance of the radiation.

When the frequency of incident radiation is above the threshold frequency the photoelectric current is directly proportional to the irradiance of the radiation.

Explaining the photoelectric effect

A beam of radiation can be regarded as a stream of individual energy bundles called **photons**. The energy E of a photon is given by

$$E = hf,$$

where h is Planck's constant and f is the frequency of the radiation.

The minimum photon energy hf_o required to release a photoelectron is called the **work function** of the surface. The work function of zinc is less than the work function of many other metals.

When N photons per second are incident on a surface the irradiance of radiation is given by

$$I = Nhf.$$

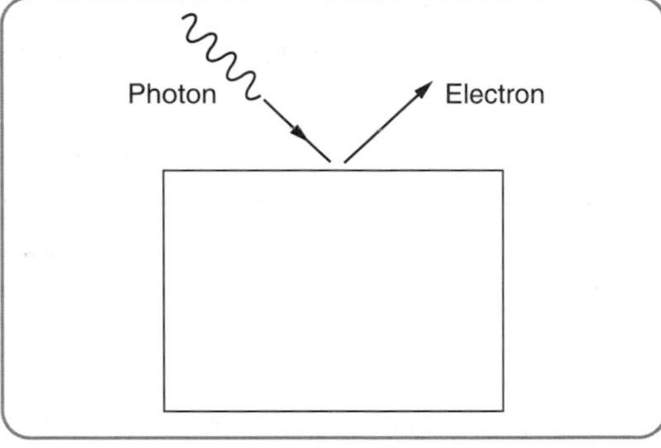

Figure 3.14

When a photon is absorbed by a material the photon is destroyed. All of its energy is given to one electron. (See Figure 3.14.)

When the photon energy is greater than the work function, some photon energy is used to release the electron from the surface. The rest of the photon energy is converted to kinetic energy of the electron.

The **maximum** kinetic energy of a photoelectron is $E_k = hf - hf_o$.

When the irradiance of the radiation is increased, more photons per second are incident on the surface. The increased number of photons release more photoelectrons.

Exercises

Exercise 39 Photoelectric effect

1 Ultraviolet radiation of frequency $1{\cdot}51 \times 10^{15}$ Hz is incident on the surface of a material of work function $9{\cdot}00 \times 10^{-19}$ J.

(a) State the minimum photon energy required to release a photoelectron.

(b) Calculate the energy of one photon of the radiation.

(c) Calculate the maximum kinetic energy of the ejected photoelectrons.

(d) Calculate the maximum velocity of the ejected photoelectrons.

2 A small lamp is used to shine radiation of frequency *f* onto a metal surface. The maximum velocity of ejected photoelectrons is *v*.

The distance between the metal surface and the lamp is doubled.

(a) Explain what effect this has on the number of photoelectrons ejected per second.

(b) Explain what happens to the maximum velocity of the photoelectrons.

3 Radiation of frequency $1{\cdot}25 \times 10^{15}$ Hz is incident on a metal surface. The work function of the metal surface is $9{\cdot}20 \times 10^{-19}$ J.

(a) Show by calculation whether photoelectrons are emitted.

(b) The irradiance of the radiation is doubled. Explain the effect this has on the number of photoelectrons ejected per second.

(c) The radiation is replaced with radiation of frequency $1{\cdot}85 \times 10^{15}$ Hz. Calculate the maximum kinetic energy of photoelectrons ejected from the surface.

3.8 Energy levels in atoms

What You Should Know

For Higher Physics you need to be able to:

- draw a diagram that represents the energy levels in a hydrogen atom
- understand and use the terms ground state, excited state, and ionisation level
- understand and use the relationships for energy and frequency of emission and absorption lines in spectra
- solve problems on energy levels in atoms.

In atoms electrons occupy discrete energy levels.

The energy levels in a hydrogen atom are represented in Figure 3.15.

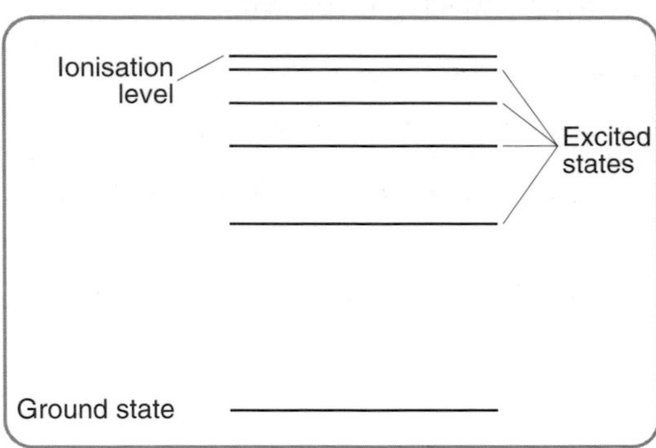

Figure 3.15

The **ground state** is the lowest electron energy level.

The **ionisation level** is the level at which the electron is released from the atom, i.e. the point at which the atom becomes a positive ion.

The **excited states** are the other energy levels between the ground state and the ionisation level.

If you are asked to draw this diagram in your Higher Physics examination, get the spacing right. The energy levels get closer together as you move from the ground state to the ionisation level.

In an excited state an electron has more energy than in the ground state, but not enough energy for ionisation.

Emission spectra

An electron in an excited state may fall to any lower energy level. This is a random process and may occur at any time.

When an electron in an excited state falls to a lower energy level a photon is created (see Figure 3.16). The electron energy is converted to the energy of the photon.

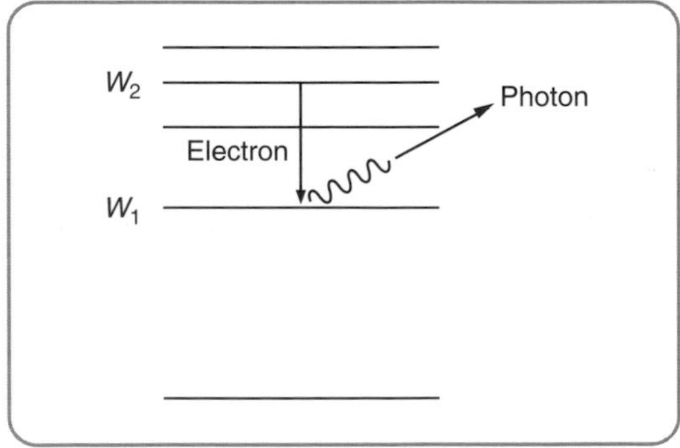

Figure 3.16

The energy of the photon is equal to the difference in energy between the states:

$$hf = W_2 - W_1.$$

Emission spectra are produced when electrons in atoms fall from excited states to lower energy levels. Lines in emission spectra are often used to identify elements.

Absorption spectra

Electromagnetic radiation incident on an atom may be absorbed by the atom.

This is only possible when the photon energy is exactly equal to the difference between two energy levels in the atom, i.e. $hf = W_2 - W_1 \Rightarrow W_2 = W_1 + hf$.

The photon is destroyed. The energy of the photon lifts the electron from one energy level to the other (see Figure 3.17).

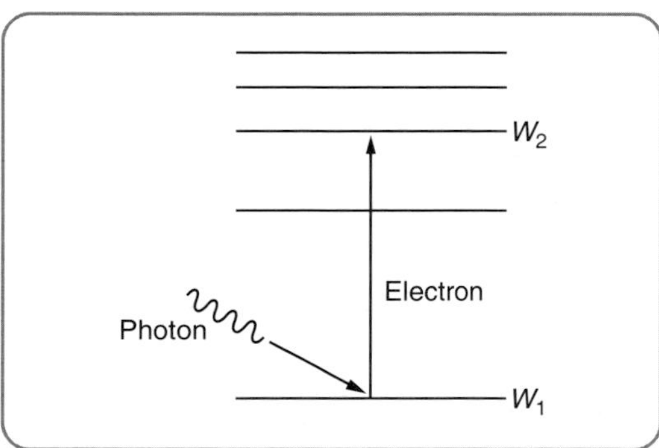

Figure 3.17

When a complete visible spectrum is incident on atoms of an element, photons with particular frequencies are absorbed. This produces dark lines in the spectrum.

The photons which are absorbed have energies exactly equal to differences between energy levels in the atoms of the element.

The spectrum of sunlight contains many dark lines. These lines are due to the absorption of photons as the sunlight passes through gases at the surface of the sun.

Exercises

Exercise 40 Energy levels in atoms

1 Atoms of a particular element have the energy levels shown below.

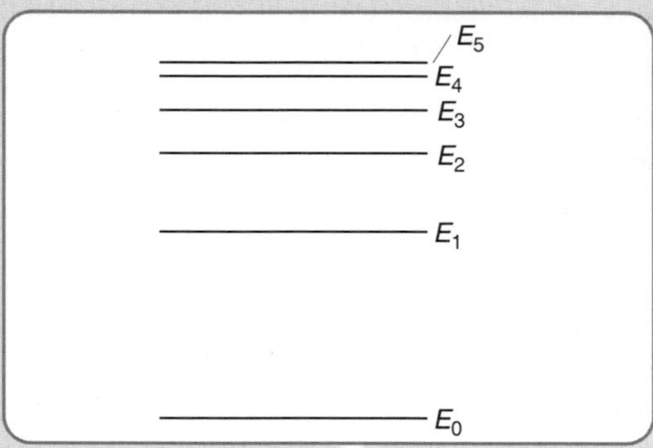

Electron transitions occur between all of these levels to produce emission lines in the spectrum of the element.

(a) How many emission lines are produced by transitions to level E_0?

(b) What is the total number of emission lines produced by transitions between these levels?

(c) Explain which transition produces photons with the shortest wavelength?

2 The following diagram shows some of the electron energy levels in atoms of an element.

A ———— 0 J

B ———— $-4{\cdot}00 \times 10^{-20}$ J

C ———— $-1{\cdot}80 \times 10^{-19}$ J

D ———— $-6{\cdot}23 \times 10^{-19}$ J

———— $-2{\cdot}21 \times 10^{-18}$ J

(a) Which level represents the ionisation level?

(b) How much energy is required to raise an electron from the ground state to level B?

(c) Calculate the frequency of a photon that could raise an electron from the ground state to level B.

(d) Calculate the wavelength of the photon created when an electron makes a transition from level C to level D.

(e) Explain whether the photon in part (d) is visible.

(f) A student compares the emission and absorption spectra of this element by superimposing same scale images of the spectra on a screen. Describe and explain what the student sees on the screen.

3.9 Lasers

What You Should Know

For Higher Physics you need to be able to:

- understand and use the terms spontaneous emission and stimulated emission
- state that in a laser the light beam gains more energy than it loses
- explain the function of the mirrors in a laser
- explain why even low power lasers are dangerous.

When an electron in an excited state falls to a lower energy level a photon is emitted. This emission is usually **spontaneous** and random, i.e. there is no external cause and it may happen at any time. There is no way of predicting when an electron will fall to a lower energy level.

When an electron in an excited state has excess energy *hf*, a photon of energy *hf* may stimulate the electron to fall to the lower energy level and emit another photon of energy *hf*. This is called **stimulated emission** (see Figure 3.18).

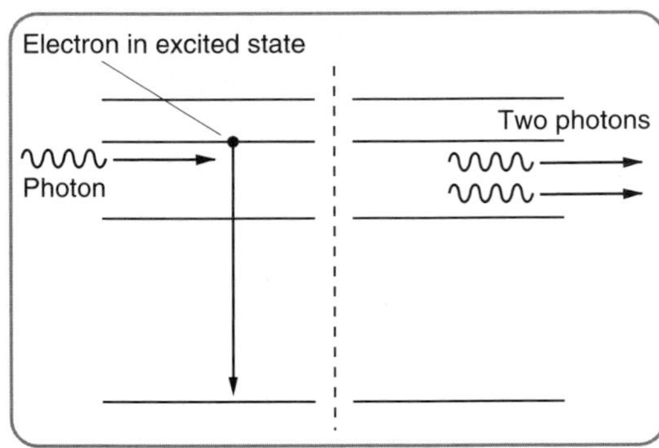

Figure 3.18

When stimulated emission occurs, the incident radiation and emitted radiation are in phase and travel in the same direction.

Conditions in a laser ensure that the light beam is amplified – it gains more energy than it loses; Hence the name: **L**ight **A**mplification by the **S**timulated **E**mission of **R**adiation: **laser**.

A laser has a mirror at either end. The function of the mirrors is to keep the photons inside the laser material so that more stimulated emissions can occur. One mirror is completely reflecting. The other is partially reflecting and allows a small proportion of light energy to pass through.

A laser beam having a power as low as 0·1 mW can cause eye damage. This is because the edges of a laser beam are very close to parallel. The irradiance of the beam is nearly constant. It does not decrease in the same way as the irradiance of light from a bulb.

Exercises

Exercise 41 Lasers

1 Explain the difference between 'spontaneous emission' and 'stimulated emission'.

2 Explain how amplification is produced in a laser.

3 A laser beam of power 0·825 W is shone onto a screen. The beam produces a bright circular spot of area 1·21 mm^2.

(a) Calculate the irradiance on the screen due to the laser beam.

(b) The distance between the laser and the screen is trebled. What effect does this have on the irradiance on the screen due to the laser beam?

3.10 Semiconductors

What You Should Know

For Higher Physics you need to be able to:

- classify materials into conductors, semiconductors and insulators
- state that doping decreases the resistance of a semiconductor
- explain how to make n-type and p-type semiconductors.

Semiconductors

Any material can be made to conduct if a high enough voltage is applied to the material. The voltage at which a material is forced to conduct is called the breakdown voltage for that material.

For voltages that are lower than breakdown voltages, materials can be classified into one of the following three categories according to their electrical properties:

- **conductors:** these have low resistance – e.g. metals, carbon (graphite)
- **insulators:** these have high resistance – e.g. plastics, rubber, asbestos, carbon (diamond)
- **semiconductors:** these have resistance values between those of good conductors and good insulators – e.g. silicon, germanium, selenium, some metal oxides.

The addition of a tiny number of impurity atoms to a pure semiconductor decreases its resistance. This process is called **doping**.

For example, pure germanium has a stable crystalline structure similar to diamond. Each germanium atom has four valence electrons which form chemical bonds with adjacent atoms.

When pure germanium is doped with atoms that have five valence electrons, four valence electrons form chemical bonds. One valence electron is 'extra'. Extra electrons can move and so carry charge within the crystal structure. The majority charge carriers are electrons. The material formed is called **n-type** germanium. (See Figure 3.19.)

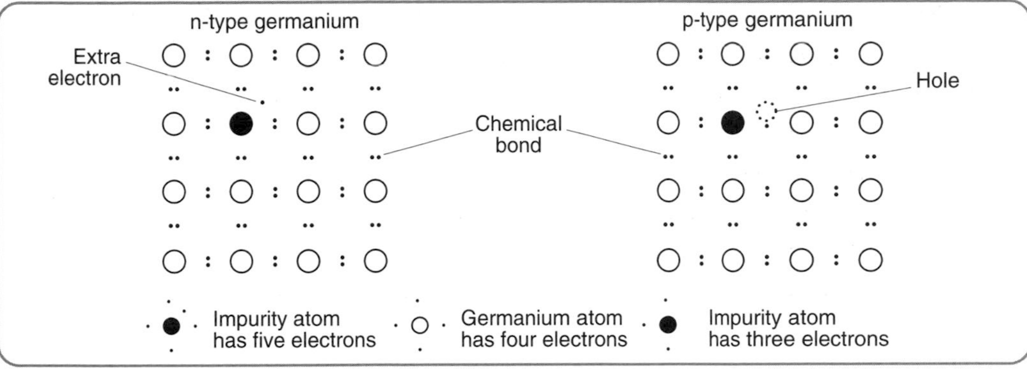

Figure 3.19

When pure germanium is doped with atoms that have three valence electrons, the valence electrons form three chemical bonds. The fourth chemical bond cannot be completed. This leaves a 'hole' in the crystal structure. The majority charge carriers are holes. Holes can be thought of as positive charge carriers moving in the opposite direction to electrons. The material formed is called **p-type** germanium.

Exercises

Exercise 42 Semiconductors

1 A neutral crystal of pure silicon is doped with a small proportion of phosphorus atoms. Phosphorus atoms have five valence electrons.
   (a) What type of semiconductor material is formed?
   (b) State the net charge on the semiconductor material formed? Explain your answer.

2 A crystal of silicon is doped with sulphur atoms. Sulphur atoms have three valence electrons. What are the majority charge carriers in the material formed?

3.11 Semiconductor devices

What You Should Know

For Higher Physics you need to be able to:

- describe processes in forward-biased and reversed-biased p–n junction diodes
- describe the uses of a photodiode in photovoltaic and photoconductive modes
- describe the structure and uses of an n-channel enhancement MOSFET.

p–n junction diode

A p–n junction diode consists of a piece of p-type semiconductor joined to a piece of n-type semiconductor. This is usually just called a diode in Higher Physics papers. Figure 3.20 shows the structure of a diode and the circuit symbol used by SQA.

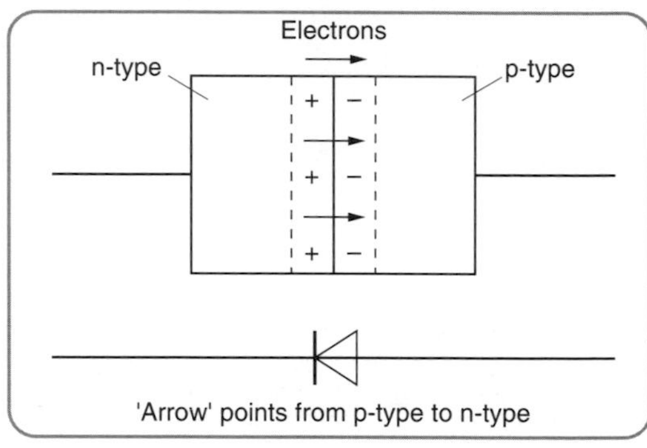

Figure 3.20

In a diode, some 'extra' electrons in the n-type cross the junction and fill holes in the p-type. As a result, the region near the junction has no charge carriers. The n-type and p-type semiconductor material becomes charged as shown.

A diode is **forward-biased** when the n-type is connected to the negative terminal of a supply and the p-type is connected to the positive terminal of the supply (see Figure 3.21).

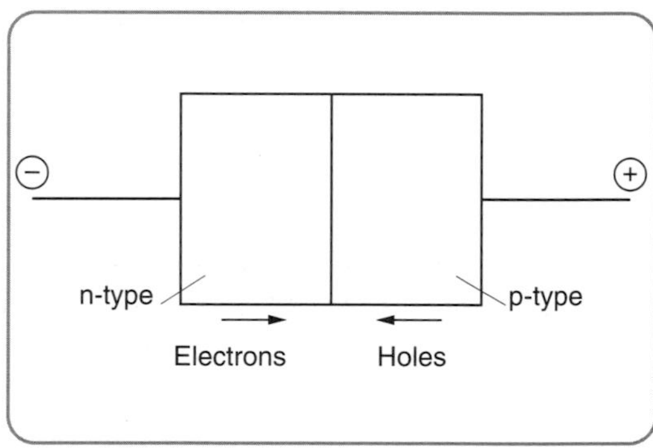

Figure 3.21

A forward-biased diode conducts.

In the n-type, electrons coming from the negative terminal replace 'extra' electrons that cross to the p-type and fill holes. In the p-type, electrons attracted towards the positive terminal are pulled out of holes that are filled. There are charge carriers throughout the diode. Electrons move towards the positive terminal and holes move towards the negative terminal.

In the junction region of a forward-biased diode, positive and negative charge carriers may recombine to emit quanta of radiation. A diode which emits quanta of radiation is called a light-emitting diode (LED).

A diode is **reverse-biased** when the n-type is connected to the positive terminal of a supply and the p-type is connected to the negative terminal of the supply (see Figure 3.22).

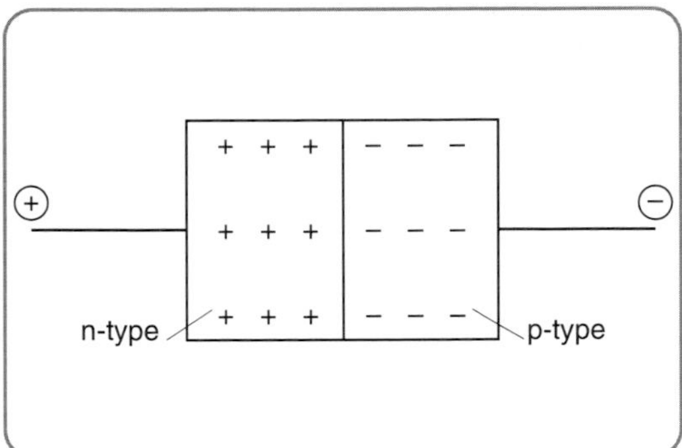

Figure 3.22

A reverse-biased diode does not conduct.

In the n-type, 'extra' electrons are attracted towards the positive terminal of the supply. This increases the size of the region with no charge carriers. In the p-type, electrons from the negative terminal of the supply fill holes. This also increases the size of the region with no charge carriers.

Remember that the 'arrow' in the diode symbol points from p-type to n-type. This will help you recognise when a diode is forward-biased and when it is reverse-biased.

As diodes conduct one way and not the other, diodes are used to convert a.c. to d.c.

Photodiode

Light incident on a p–n junction may produce positive and negative charges. A diode in which this happens is called a photodiode.

A photodiode may be used to supply power to a circuit. When it is used in this way the diode is in **photovoltaic** mode. (See Figure 3.23.) Light energy incident on the p–n junction is the source of electrical energy in the circuit.

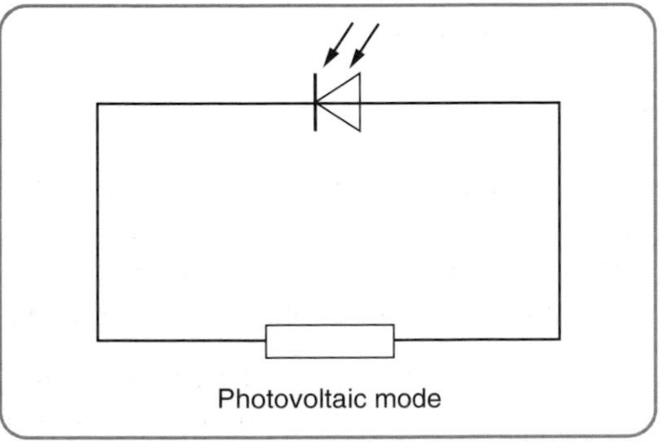

Figure 3.23

A **reverse-biased** photodiode may be used as a light sensor. When it is used in this way the diode is in **photoconductive** mode. (See Figure 3.24.) In photoconductive mode a photodiode acts as a light-dependent resistor (LDR).

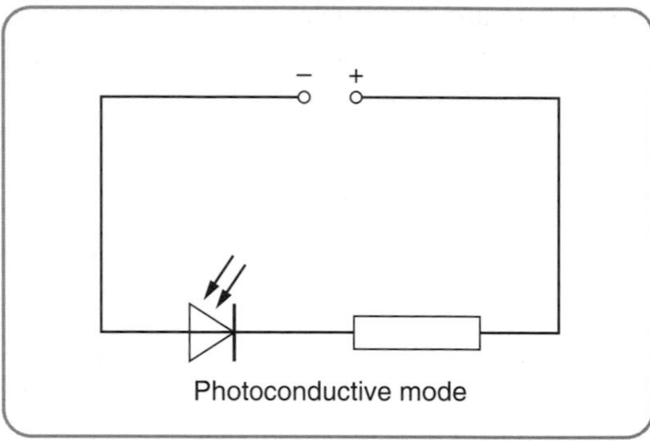

Figure 3.24

The current in the diode is called leakage current. The leakage current is directly proportional to the irradiance of the light incident on the photodiode. For voltages lower than the breakdown voltage (the voltage at which the diode is forced to conduct) the current is fairly independent of the reverse-biasing voltage.

Reverse-biased photodiodes respond very quickly to changes in light level. This enables reverse-biased photodiodes to be used as very fast switches in telecommunications equipment and in barcode scanners.

n-channel enhancement MOSFET

A MOSFET is a **M**etal **O**xide **S**emiconductor **F**ield **E**ffect **T**ransistor.

The structure and circuit symbol of an n-channel enhancement MOSFET are shown in Figure 3.25. Note that the back electrode is connected to the source.

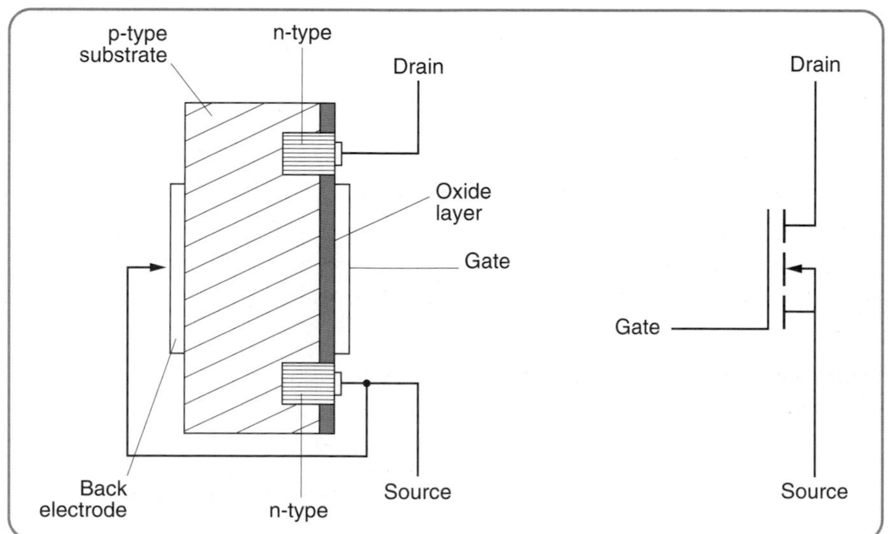

Figure 3.25

You need to learn these for your Higher Physics.

You also need to be able to explain the electrical ON and OFF states of a MOSFET.

When an n-channel enhancement MOSFET is connected in a circuit, the drain is made more positive than the source. As the back electrode is connected to the source the junction between the substrate and the source is unbiased. The junction between the drain and the substrate is reverse-biased. This reverse bias and the high resistance of the substrate prevent the flow of charge. The MOSFET is **off**.

To switch on the MOSFET a positive voltage of over 2 V is applied to the gate. This sets up an electric field between the gate and the back electrode. Electrons in the p-type substrate gather in a layer beneath the gate and form a channel which enables charge to flow from the source to the drain. The MOSFET is **on**.

When the MOSFET is on, the gate voltage controls the current between the source and the drain.

An n-channel enhancement MOSFET can be used as an amplifier.

Exercises

Exercise 43 Semiconductor devices

1 The circuit below shows a photodiode connected so that it can be used as a light sensor.

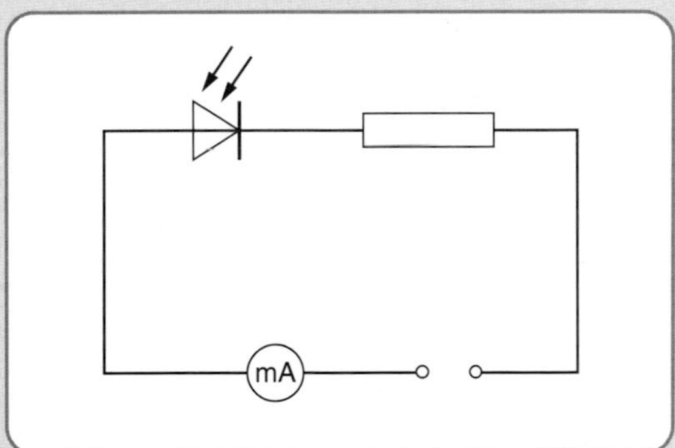

(a) Copy the circuit and add '+' and '–' signs to show how the supply should be connected so that the photodiode is correctly biased.

(b) The circuit is in a darkened laboratory. A small lamp at a distance of 1·50 m from the photodiode is switched on. The reading on the meter is 0.280 mA. The lamp is moved to a distance of 2·50 m from the photodiode. Calculate the new reading on the meter.

(c) Explain your answer to part (b) in terms of the action of photons at the p-n junction.

***Exercises** continued* ➢

Exercises continued

2 The LED in the circuit below emits photons of wavelength 595 nm in air.

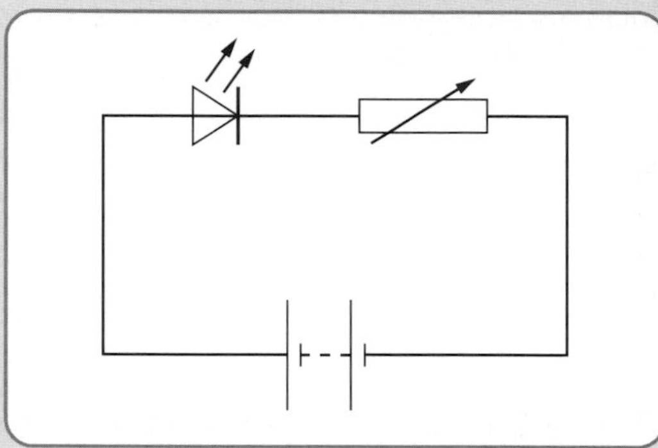

(a) Calculate the energy of a photon of light from this LED.

(b) Calculate the minimum p.d. across the p–n junction when it emits photons.

(c) Explain how light is produced at the p–n junction in terms of majority charge carriers.

3 A sinusoidal a.c. supply is connected to the arrangement of four diodes shown below.

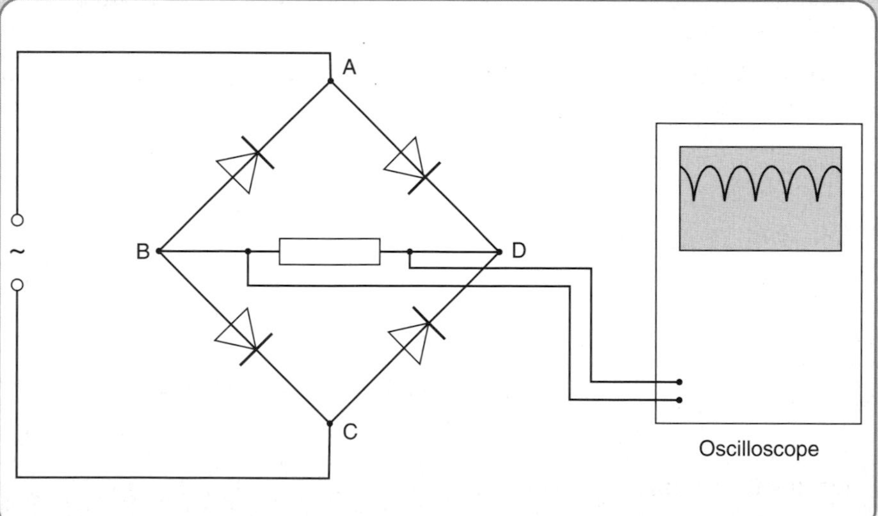

The trace on the oscilloscope screen shows that the current in the resistor is always in the same direction.

(a) Explain this observation in terms of the bias of the diodes.

(b) Suggest a method for reducing the variation of the p.d. across the resistor.

Exercises continued ➢

Exercises continued

4 A student sets up the circuit below to investigate the behaviour of an n-channel enhancement MOSFET.

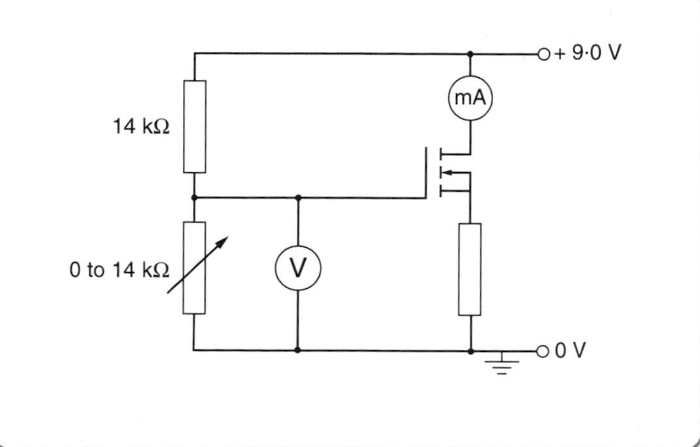

The variable resistor is originally set at 0 Ω.

Describe and explain what happens to the readings on the meters as the resistance of the variable resistor is gradually increased from 0 Ω to 14 kΩ.

Significant resistance values and voltmeter readings should be specified. Current values are not required.

3.12 Nuclear reactions

What You Should Know

For Higher Physics you need to be able to:

- describe how Rutherford discovered the nucleus
- identify the processes alpha decay, beta decay and gamma decay of radionuclides
- describe fission and fusion reactions
- use $E = mc^2$ to explain why the products of fission and fusion have large amounts of kinetic energy
- solve problems on nuclear reactions.

Rutherford model of the atom

When alpha particles are fired in a vacuum at a thin sheet of gold, most of the alpha particles pass straight through. This suggests that most of a gold atom is empty space.

A small number of alpha particles bounce back in the direction from which they came. This suggests that some alpha particles collide with an object that has very large mass and the same charge as an alpha particle.

From these observations Rutherford concluded the following:

- Most of the mass of an atom is concentrated in a tiny massive core – the nucleus.
- The diameter of the nucleus is much smaller than the diameter of the atom.

In the Rutherford model of the atom, electrons orbit the nucleus.

Further research established that a nucleus is made up of protons and neutrons. The following symbols are used by SQA to specify individual nuclei (nuclei is the plural of nucleus).

$${}^{A}_{Z}\mathrm{X}$$

A is the mass number = (number of protons + number of neutrons).
Z is the atomic number = number of protons.
X is the chemical symbol of the element.

Radioactive decay

The nuclei of certain atoms decay spontaneously. This process is random, like the spontaneous emission of photons when electrons fall from excited states to lower energy levels. There is no way to predict when an individual nucleus will decay.

Both mass number *A* and atomic number *Z* are conserved in radioactive decay.

There are three types of spontaneous radioactive decay and you have to be able to recognise each of these for Higher Physics.

In **alpha decay** a helium nucleus is ejected from a nucleus. *A* falls by 4 and *Z* decreases by 2, e.g.

$${}^{223}_{89}\mathrm{Ac} \rightarrow {}^{219}_{87}\mathrm{Fr} + {}^{4}_{2}\mathrm{He}.$$

In **beta decay** an electron is ejected from a nucleus. *A* does not change and *Z* increases by 1, e.g.

$${}^{215}_{84}\mathrm{Po} \rightarrow {}^{215}_{85}\mathrm{At} + {}^{0}_{-1}\mathrm{e}.$$

In **gamma decay** a high energy photon is ejected from a nucleus. Neither *A* nor *Z* changes, e.g.

$${}^{226}_{88}\mathrm{Ra} \rightarrow {}^{226}_{88}\mathrm{Ra} + \text{photon}.$$

Fission

In fission a nucleus of large mass number splits into two nuclei of smaller mass number. Usually a small number of neutrons are also released. Natural fission is usually **spontaneous**.

Fission may also be **induced** by bombarding a large nucleus with neutrons. For induced fission, neutrons have to be travelling slowly enough for the nucleus to capture a neutron. The nucleus becomes unstable and fission occurs.

Whether spontaneous or induced the products of a fission reaction have smaller mass than the reactants. The mass which is lost is converted to kinetic energy of the products.

The mass lost and energy released are linked by the relationship $E = mc^2$, where c is the speed of light in a vacuum.

As c is a very large number, a tiny amount of mass is equivalent to a huge amount of energy.

Fusion

In fusion, two nuclei of small mass number combine to form a nucleus of larger mass number.

The products of a fusion reaction also have smaller mass than the reactants. The mass which is lost is converted to kinetic energy of the products.

As above, the mass lost and energy released are linked by the relationship $E = mc^2$.

Exercises

Exercise 44 Nuclear reactions

1 Why was Rutherford's alpha scattering experiment carried out in a vacuum?

2 Part of a radioactive decay series is shown below.

$$^{238}_{92}\text{U} \rightarrow {}^{234}_{90}\text{Th} \rightarrow \ldots \rightarrow {}^{226}_{88}\text{Ra} \rightarrow {}^{222}_{86}\text{Rn}$$

(a) Calculate the number of alpha particles emitted.

(b) Calculate the number of beta particles emitted.

(c) Calculate the number of gamma photons emitted.

3 The incomplete relationship below represents a reaction that takes place in the core of a nuclear reactor.

$$^{235}_{92}\text{U} + {}^{1}_{0}\text{n} \rightarrow {}^{139}_{57}\text{La} + {}^{95}_{41}\text{Nb} + \ldots$$

The missing products are neutrons and electrons.

(a) How many protons are there in the lanthanum (La) nucleus?

(b) How many neutrons are there in the niobium (Nb) nucleus?

(c) How many neutrons are released in this reaction? You must justify your answer by calculation.

(d) How many electrons are released in this reaction? You must justify your answer by calculation.

4 The relationship below represents a reaction which takes place in a nuclear reactor.

$$^{235}_{92}\text{U} + {}^{1}_{0}\text{n} \rightarrow {}^{144}_{58}\text{Ce} + {}^{89}_{37}\text{Rb} + 3\,{}^{1}_{0}\text{n} + 3\,{}^{0}_{-1}\text{e}$$

(a) State the type of fission reaction represented by this relationship. Explain your answer.

Exercises *continued* ➢

Exercises *continued*

(b) Calculate the energy released in the reaction. The masses of particles involved are:

neutron	= $1{\cdot}674\ 93 \times 10^{-27}$ kg	rubidium-89	= $1{\cdot}476\ 07 \times 10^{-25}$ kg
uranium-235	= $3{\cdot}902\ 15 \times 10^{-25}$ kg	cerium-144	= $2{\cdot}389\ 21 \times 10^{-25}$ kg
electron	= negligible		

(c) Explain how neutrons released by this reaction can cause further nuclear reactions.

5 The following reaction takes place inside a fusion reactor.

$$^{2}_{1}\text{H} + ^{3}_{1}\text{H} \rightarrow ^{4}_{2}\text{He} + ^{1}_{0}\text{n}$$

The masses of the particles involved in the reaction are:

$$^{2}_{1}\text{H} : 3{\cdot}344\ 49 \times 10^{-27}\,\text{kg} \quad ^{3}_{1}\text{H} : 5{\cdot}008\ 26 \times 10^{-27}\,\text{kg}$$

$$^{4}_{2}\text{He} : 6{\cdot}646\ 47 \times 10^{-27}\,\text{kg} \quad ^{1}_{0}\text{n} : 1{\cdot}674\ 93 \times 10^{-27}\,\text{kg}$$

(a) Calculate the energy released in the reaction.

(b) Calculate the number of reactions required each second to generate 4·5 MW of heat energy.

3.13 Dosimetry and safety

What You Should Know

For Higher Physics you need to be able to:

- define the activity of a radioactive source
- understand and use the quantities absorbed dose, tissue weighting factor, equivalent dose, equivalent dose rate and effective dose
- describe factors affecting background radiation
- describe how to reduce equivalent dose rate
- sketch a graph to show how the irradiance due to a beam of gamma rays varies with the thickness of absorber in the path of the beam
- describe how to measure half-value thickness
- solve problems on dosimetry and safety.

The activity A of a radioactive source is the average number of nuclei decaying per unit time.

$A = \dfrac{N}{t}$, where N is the number of nuclei decaying in time t.

The SI unit of activity is the becquerel (Bq). 1 Bq = 1 decay per second.

Exposure to radiation and risk

When tissue is exposed to radiation, the **absorbed dose** D is the energy absorbed per unit mass of the tissue. The unit of absorbed dose is the gray (Gy). 1 Gy = 1 J kg^{-1}.

$$D = \frac{E}{m}$$

The risk of biological harm to tissue depends on the:

- absorbed dose
- kind of radiation
- body organs or tissue exposed.

A radiation weighting factor w_R is given to each radiation. This factor is a measure of the biological effect of the radiation.

Equivalent dose H is the product of absorbed dose and radiation weighting factor. The unit of equivalent dose is the sievert (Sv).

$$H = Dw_R$$

Equivalent dose rate $\dot{H}$ is the equivalent dose per unit time. $\dot{H} = \frac{H}{t}$.

Some tissues are more likely than others to be damaged by exposure to ionising radiation.

Effective dose takes into account the different susceptibilities to harm of body tissues. Effective dose indicates the overall risk to health from exposure to ionising radiation.

Exercises

Exercise 45 Dosimetry

1 A student uses a Geiger counter to measure the activity of a source of beta particles. The Geiger counter registers 12 000 counts in 5 minutes. Calculate the activity of the source.

2 The activity of 100 g of a pure radioactive mineral is $2{\cdot}3 \times 10^4$ decays per second. 15% of the mass of a 2·4 kg rock is the mineral. The remaining 85% of mass of the rock is not radioactive. Calculate the activity of the rock.

3 A technician works an average of 125 hours each month in an area exposed to a neutron beam. In a 12 month period the total equivalent dose received by the technician is $3{\cdot}75 \times 10^{-2}$ Sv.

(a) Calculate the hourly equivalent dose rate received by the technician.

(b) The radiation weighting factor of the neutron beam is 4. Calculate the average absorbed dose received by the technician each month.

***Exercises** continued* ➢

Exercises continued

4 The table below gives information on the exposure of body tissue to three types of radiation.

Radiation	*Mass of tissue/g*	*Energy of radiation absorbed/mJ*	*Radiation weighting factor*
Gamma rays	150	46	1
Fast neutrons	320	12	8
Alpha particles	260	6·5	20

(a) The exposure to which of these types of radiation carries the greatest risk of biological harm? You must justify your answer by calculation.

(b) What other factor must be taken into account to indicate the risk to health from exposure to these radiations.

Background radiation

Every day we are all exposed to small amounts of radiation called background radiation. The following contribute to background radiation.

Cosmic rays: The Earth is constantly bombarded by high energy particles from space. Most cosmic rays come from the sun. The atmosphere and the Earth's magnetic field give us some protection but some of the cosmic rays get through to the Earth's surface.

Radioactive elements: Many naturally occurring elements are radioactive. The level of background radiation in any area depends on the number and type of radioactive elements in the rocks, soil, water and air in that area.

Radon gas: Radon is part of the radioactive decay series of uranium. Radon accounts for more than half of background radiation. Radon gas is dense and may accumulate in the basement of buildings.

Human body: We are all slightly radioactive because we eat, drink and breathe radioactive elements in our surroundings. For example, the level of the radioactive isotope carbon-14 in our bodies is constant while we are alive.

The level of background radiation varies enormously around the world. In the UK we are each exposed to an effective dose of around 2 mSv per year.

Radiation and safety

To reduce the risks to health, the government has set an annual effective dose limit of 1 mSv for the exposure of the general public to radiation. A higher limit of 20 mSv per year has been set for people who work in industries that use radioactive materials.

Equivalent dose rate from a radioactive source is reduced with increasing distance from the source. Equivalent dose rate is also reduced by **shielding**, i.e. the placing of absorbing material in the path of radiation. The reduction in equivalent dose rate depends on the absorber. For example lead is more effective than aluminium in reducing equivalent dose rate. Thicker absorbers reduce equivalent dose rate more than thinner absorbers made of the same material.

The half-value thickness of an absorber for a particular radiation is the thickness that absorbs half of the radiation.

Experiments

Measuring half-value thickness of an absorber

The apparatus in Figure 3.26 is used to measure the half-value thickness of an absorbing material.

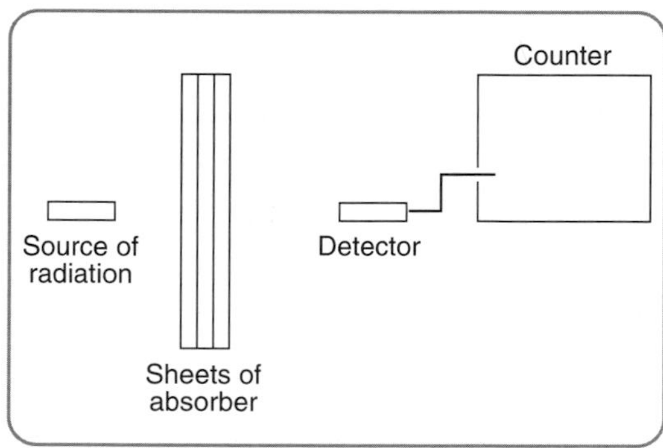

Figure 3.26

The level of background radiation is measured using the detector and counter.

The source is set at a fixed distance in front of the detector. The count rate is measured.

A measured thickness of absorbing material is placed between the source and detector. The count rate is again measured. The procedure is repeated with additional measured thicknesses of absorber between the source and detector. Corrected count rates are found by subtracting background count rate from each measured value of count rate.

A graph of corrected count rate against thickness of absorber is drawn. The graph in Figure 3.27 is obtained.

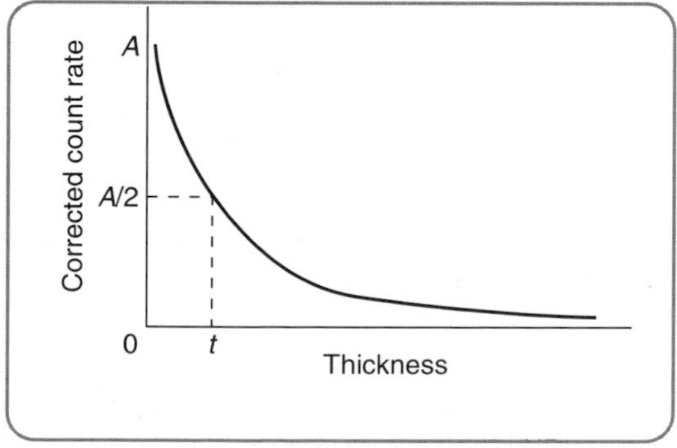

Figure 3.27

The graph is used to find the thickness of absorber for which the corrected count rate is half the initial value. Thicknesses of absorber required to halve four or five other values of count rate are found from the graph. The average value calculated is the half-value thickness of the material for the radiation.

Exercises

Exercise 46 Radiation and safety

1 In an experiment a student uses a radioactive source to irradiate a 0·65 kg sample of tissue. The absorbed dose is $4{\cdot}8 \times 10^{-5}$ Gy.

(a) Calculate the energy absorbed by the tissue.

(b) The student repeats the experiment with a 120 mm thick sheet of lead placed between the source and the tissue. The absorbed dose in this experiment is $3{\cdot}0 \times 10^{-6}$ Gy. Calculate the half-value thickness of the lead for this radiation.

2 A Geiger counter placed 4·5 m from a source of gamma rays registers a count rate of 50 counts per minute. Calculate the count rate when the counter is moved to a distance of 1·5 m from the source.

Chapter 4

COURSE SKILLS

4.1 Units, prefixes and scientific notation

What You Should Know

For Higher Physics you need to be able to:

- use SI units of all physical quantities included in the course
- give answers to calculations to an appropriate number of figures
- use scientific notation
- understand and use the prefixes p, n, μ, m, k, M, G.

Units

You have to be able to use the SI units of all physical quantities included in the Higher Physics course. A full list of quantities is included in the table at the end of this section.

Get into the habit of using correct SI units every time you tackle a numerical problem. This habit will save you time in the examination and will help you get units correct without having to think.

In Higher Physics, there is usually 1/2 mark for the correct unit in the **final answer** to a numerical problem. It is not much but these half marks accumulate into enough marks to make the difference between passing and failing.

You do not have to write units in the middle of a calculation. You **must** include the correct unit in your statement of the final answer.

Significant figures

The number of significant figures in your **final answer** to a numerical question (or part of a question) should be the same as the **minimum** number of significant figures in the data that you use to work out the answer.

Do not round intermediate values to the number of significant figures for the question. Where possible keep one additional figure in intermediate values and **round when you write your final answer**. This method is used in the examples and the solutions to the exercises in this book.

Be careful about significant figures when you use data from the data sheet at the front of the examination paper. For example, the value of *g* is usually quoted to only **two** figures. If you use *g* to work out an answer to a question where the data are quoted to three figures, two is the correct number of significant figures for your answer.

Having too many figures in final answers will cost you marks – do not let this be the reason why you do not pass your Higher!!

Scientific notation

Some data in Higher Physics are very large or very small numbers. You have to understand and use the notation for these numbers, e.g.

$$\text{Planck's constant} = 6{\cdot}63 \times 10^{-34}\ \text{J s}.$$

That you have reached this far in the book means you probably already have this skill. If not, learn it quickly.

Prefixes

Table 4.1 includes all the prefixes that you need to understand for Higher Physics. Learn them and get into the habit of using them.

Table 4.1 Prefixes

Prefix	*Short for*	*Means*	*Prefix*	*Short for*	*Means*
m	milli	$\times 10^{-3}$	k	kilo	$\times 10^{3}$
µ	micro	$\times 10^{-6}$	M	mega	$\times 10^{6}$
n	nano	$\times 10^{-9}$	G	giga	$\times 10^{9}$
p	pico	$\times 10^{-12}$			

Note: The kilogram is the SI unit for mass – **do not** change kilograms to grams.

Exercises

Exercise 47 Units, prefixes and scientific notation

1 Convert the following quantities to scientific notation.
  (a) 40 000 V
  (b) 0·000 036 5 W m^{-2}
  (c) 200 pF
  (d) 4500 km
  (e) 5·0 GHz
  (f) 1·25 µA.

2 Convert the following quantities to decimal numbers with the correct SI unit.
  (a) $9{\cdot}0 \times 10^{-2}$ kg m^{-3}
  (b) 724 mm
  (c) $0{\cdot}098 \times 10^{2}$ m s^{-2}
  (d) 8·8 MJ
  (e) $3{\cdot}16 \times 10^{-5}$ Sv.

Physical quantities in Higher Physics

Table 4.2 Physical quantities

Physical quantity	*Symbol*	*Unit*	*Abbreviation*
Distance, displacement	s or d	metre	m
Height, depth	h	metre	m
Time	t	second	s
Speed, velocity, final velocity	v	metre per second	m s^{-1}
Initial velocity	u	metre per second	m s^{-1}
Change of velocity	Δv	metre per second	m s^{-1}
Average velocity	$\bar{v}$	metre per second	m s^{-1}
Acceleration	a	metre per second per second	m s^{-2}
Acceleration due to gravity	g	metre per second per second	m s^{-2}
Gravitational field strength	g	newton per kilogram	N kg^{-1}
Mass	m	kilogram	kg
Force, upthrust, tension	F	newton	N
Weight	W	newton	N
Energy	E	joule	J
Work done	W or E_w	joule	J
Potential energy	E_p	joule	J
Kinetic energy	E_k	joule	J
Power	P	watt	W
Momentum	p	kilogram metre per second	kg m s^{-1}
Impulse	(Δp)	newton second	N s
Volume	V	cubic metre	m^3
Density	ρ	kilogram per cubic metre	kg m^{-3}
Area	A	square metre	m^2
Pressure	P or p	pascal	Pa
Temperature	T	degree celsius or kelvin	°C or K
Electric charge	Q	coulomb	C
Electric current	I	ampere	A
Peak current	I_{peak}	ampere	A
Root mean square current	$I_{r.m.s.}$	ampere	A
Voltage, potential difference	V	volt	V
Electromotive force	E or ε	volt	V

Table 4.2 Physical quantities (*continued*)

Physical quantity	*Symbol*	*Unit*	*Abbreviation*
Peak voltage	V_{peak}	volt	V
Root mean square voltage	$V_{r.m.s.}$	volt	V
Input voltage	V_1 or V_2	volt	V
Output voltage	V_o	volt	V
Internal resistance	r	ohm	Ω
Resistance	R	ohm	Ω
Feedback resistance	R_f	ohm	Ω
Capacitance	C	farad	F
Voltage gain	A_o	–	–
Period	T	second	s
Frequency	f	hertz	Hz
Wavelength	λ	metre	m
Angle	θ	degree	°
Critical angle	θ_c	degree	°
Refractive index	n	–	–
Irradiance	I	watt per square metre	$W\,m^{-2}$
Planck's constant	h	joule second	J s
Number of photons	N	–	–
Threshold frequency	f_o	hertz	Hz
Energy level	$W_1, W_2, \ldots$	joule	J
Speed of light in a vacuum	c	metre per second	$m\,s^{-1}$
Activity	A	becquerel	Bq
Count rate	–	counts per second	–
Number of nuclei decaying	N	–	–
Absorbed dose	D	gray	Gy
Radiation weighting factor	w_R	–	–
Equivalent dose	H	sievert	Sv
Equivalent dose rate	$\dot{H}$	sievert per second	$Sv\,s^{-1}$
Effective dose	H	sievert	Sv
Half-life	$t_{1/2}$	second	s
Half-value thickness	$T_{1/2}$	metre	m

4.2 Uncertainties

What You Should Know

For Higher Physics you need to be able to:

- express uncertainties in absolute and percentage forms
- distinguish between random and systematic uncertainties
- estimate scale-reading uncertainties for analogue and digital scales
- calculate a mean and the approximate uncertainty in the mean.

There is **always** uncertainty in the measurement of a physical quantity.

Experimental measurements are stated in the form

$$\text{measurement} \pm \text{absolute uncertainty.}$$

For example, the output voltage of an op-amp

$$V_o = 1{\cdot}0 \text{ V} \pm 0{\cdot}1 \text{ V} \quad \textbf{or} \quad V_o = (1{\cdot}0 \pm 0{\cdot}1) \text{ V.}$$

Measurements are also stated in the form

$$\text{measurement} \pm \text{percentage uncertainty.}$$

For example, the output voltage above of the op-amp is $V_o = (1{\cdot}0 \pm 10\%)$ V.

$$\text{percentage uncertainty} = \frac{\text{absolute uncertainty}}{\text{measurement}} \times 100\%$$

In Higher Physics you could be given a set of measurements and their uncertainties and asked to identify the measurement with the biggest percentage uncertainty.

In an experiment the largest percentage uncertainty in any one measurement is often a good estimate of the percentage uncertainty in the final numerical result.

Uncertainties are reduced by careful experimental design and by experimenters taking care over measurements. Valid experimental results rely on uncertainties being reduced as far as possible.

Random uncertainties

The effects of random uncertainties are not predictable and cannot be completely eliminated from an experiment.

Repeating measurements and calculating a mean reduces uncertainty and leads to more accurate measurement. The mean is the best estimate of the true value of the quantity being measured.

The mean of *n* measurements of a quantity is the sum of the measurements divided by *n*.

$$\text{mean value of X} = \frac{X_1 + X_2 + X_3 + \ldots + X_n}{n}.$$

The uncertainty in the mean is estimated by dividing the difference between the largest measurement and the smallest measurement by *n*.

$$\text{uncertainty in mean} = \frac{\text{maximum value} - \text{minimum value}}{n}$$

The above relationship is an approximation which is reasonably accurate when at least five readings have been taken.

A scale-reading uncertainty is a measure of how well an instrument scale can be read. For analogue scales the scale-reading uncertainty is usually ± half of the smallest scale division. For digital scales the scale-reading uncertainty is ± 1 in the last digit.

Systematic uncertainties

Systematic uncertainties have consistent effects on quantities being measured.

Effects of systematic uncertainties can be identified by plotting a graph. For example, in an experiment you may expect to get a straight line graph through the origin. If you get a straight line graph that does not pass through the origin it is likely that there is a systematic effect.

Where there is a systematic effect the mean is offset from the true value.

Identifying systematic effects is often an important part of evaluating experiments.

Exercises

Exercise 48 Uncertainties

1 In an experiment to measure the specific heat capacity of a liquid a student records the following measurements.

mass of liquid = 1·15 kg ± 0·01 kg
heater voltage = 9·41 V ± 0·02 V
heater current = 6·37 A ± 0·02 A
time = 1200 s ± 2 s
temperature rise = 25 °C ± 1 °C

(a) Use $E_h = cm\Delta T$ to calculate the specific heat capacity of the liquid.

(b) State the result of the student's experiment in the form

measurement ± absolute uncertainty.

2 In an experiment to measure the absolute refractive index of a rectangular block of plastic a student records the following measurements.

Angle in air θ_1/°	20·0	30·0	40·0	50·0	60·0	70·0
Angle in plastic θ_2/°	13·5	19·5	25·5	31·0	36·0	39·5

Uncertainty in measurement of each angle = ± 0·5°.

(a) Calculate the mean refractive index of the plastic.

(b) Calculate the random uncertainty in the mean.

(c) Which pair of measurements on its own is likely to give the most accurate value of the refractive index of the plastic? Justify your answer.

Chapter 5

SOLUTIONS

5.1 Mechanics

Exercises

Exercise 1 Scalars and vectors

1 All scalars – lists (b), (d), (f) and (h).

2 All vectors – lists (c), (e) and (g).

Exercises

Exercise 2 Adding vectors

1

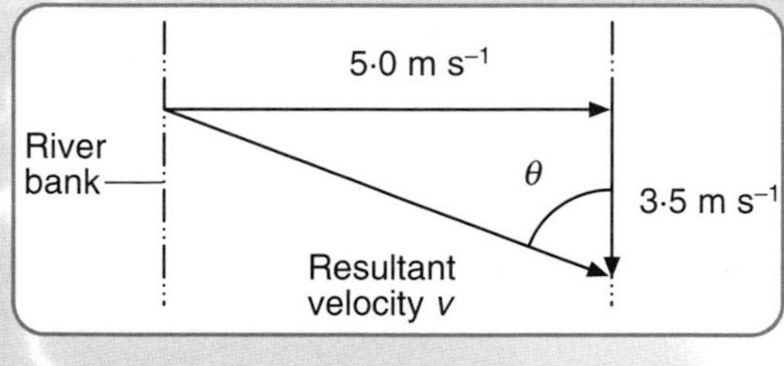

By Pythagoras: $v^2 = 5{\cdot}0^2 + 3{\cdot}5^2$

$= 37{\cdot}25$

$\Rightarrow v = 6{\cdot}1 \text{ m s}^{-1}$

$\tan\theta = \dfrac{5{\cdot}0}{3{\cdot}5} \Rightarrow \theta = 55°$ to the river bank

$\Rightarrow v = 6{\cdot}1 \text{ m s}^{-1}$ at 55° to the river bank

2

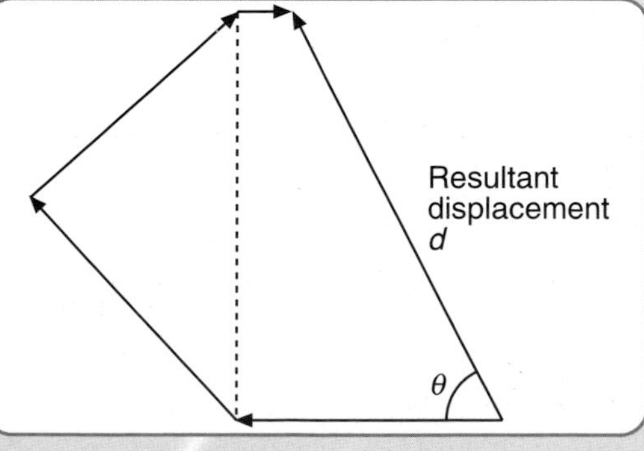

Your scale drawing should look like this:

$d = 682$ m, $\theta = 68{\cdot}9°$

resultant = 680 m 21° West of North

(**Note:** Your answer is correct if the shape of your diagram looks right **and** you are within ± 10 m of 680 – **and** you are within ± 2° of 21°.)

Exercises

Exercise 3 Components of vectors

1 (a) Answer should have **three** figures, because the data in the question have a minimum of three significant figures.

(b) Horizontal component = $v \cos \theta = 40 \cos 50° = 25{\cdot}71 = 25{\cdot}7 \text{ m s}^{-1}$.

(c) Vertical component = $v \sin \theta = 40 \sin 50° = 30{\cdot}64 = 30{\cdot}6 \text{ m s}^{-1}$.

2 (a) Component down the slope = $mg \sin \theta = 4{\cdot}6 \times 9{\cdot}8 \times \sin 37° = 27{\cdot}1 = 27$ N.

(b) Component into the slope = $mg \cos \theta = 4{\cdot}6 \times 9{\cdot}8 \times \cos 37° = 36{\cdot}0 = 36$ N.

Exercises

Exercise 4 Acceleration

1 $a = \dfrac{v - u}{t} = \dfrac{25 - 0}{10} = 2{\cdot}5 \text{ m s}^{-2}$

2 $a = \dfrac{v - u}{t} = \dfrac{0 - 10}{4} = -2{\cdot}5 \text{ m s}^{-2}$ **Note:** *Substitution must be correct!!*

3 $a = \dfrac{v - u}{t} \Rightarrow 0{\cdot}2 = \dfrac{v - 3{\cdot}4}{3} \Rightarrow v = 4{\cdot}0 \text{ m s}^{-1}$

Exercises

Exercise 5 Measuring acceleration

1 (a) $u = \dfrac{s}{t} = \dfrac{0{\cdot}15}{0{\cdot}30} = 0{\cdot}5 \text{ m s}^{-1}$

$v = \dfrac{s}{t} = \dfrac{0{\cdot}15}{0{\cdot}15} = 1{\cdot}0 \text{ m s}^{-1}$

$a = \dfrac{v - u}{t} = \dfrac{1 - 0{\cdot}5}{2{\cdot}0} = 0{\cdot}25 \text{ m s}^{-2}$

(b) There should be two significant figures in the final answer.

Exercises

Exercise 6 Graphs of motion

1 (a)

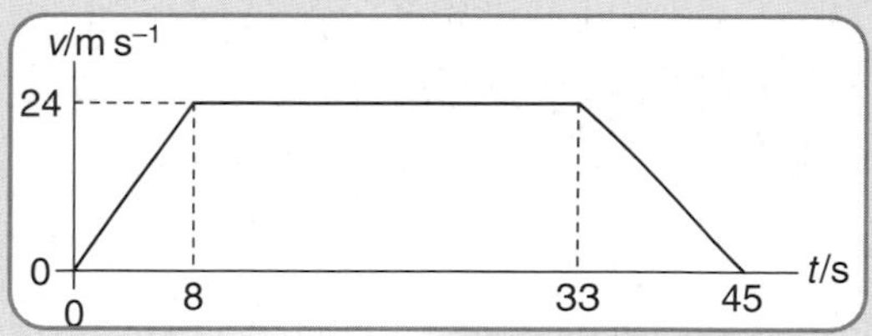

(b)

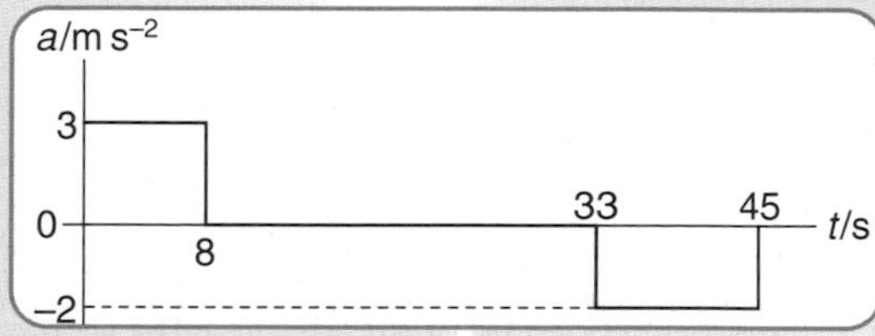

Calculations for a: first 8 seconds: $a = \dfrac{24 - 0}{8} = 3 \text{ m s}^{-2}$

next 25 seconds: $a = 0$ as v–t graph is parallel to time axis

last 12 seconds: $a = \dfrac{0 - 24}{12} = -2 \text{ m s}^{-2}$

(c) total distance travelled = area under v–t graph

$= (½ \times 8 \times 24) + (25 \times 24) + (½ \times 12 \times 24) = 840$ m

average speed $= \dfrac{\text{total distance}}{\text{total time}} = \dfrac{840}{45} = 18{\cdot}67 = 19 \text{ m s}^{-1}$

Why is the final answer rounded to two figures?

2 (a) Ball is moving downwards for all of the parts of the graph below the time axis. OA represents the initial movement of the ball – this is downwards. For OA the velocity is negative; this means that when v is negative the ball is moving down.

(b) All of the parts of the graph above the time axis (i.e. the parts where v is positive).

(c) AB and DE represent the bounces (the direction changes from down to up).

(d) The ball is stationary at C. $v = 0$ (this is the highest point after the first bounce).

(e) AB and DE represent constant positive accelerations.

(f) OA and BCD (and the bit after E) represent constant negative accelerations.

Exercises *continued* ➢

Exercises continued

(g)

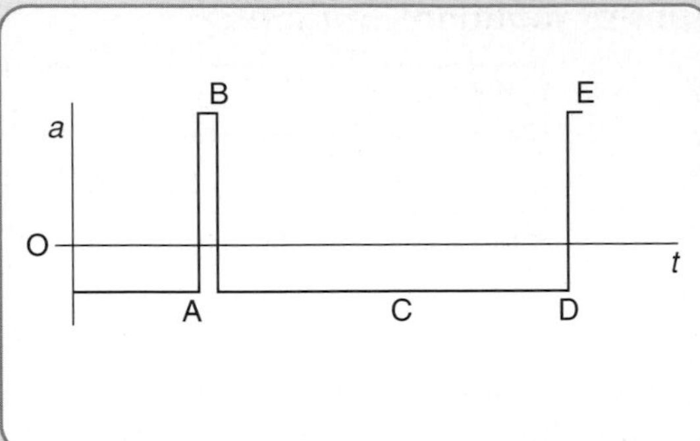

Exercises

Exercise 7 Deriving the equations of motion

See the derivations at the start of Section 1.4.

Exercises

Exercise 8 Using the equations of motion

1 $u = 6{\cdot}4\ \text{m s}^{-1}$ — ***Collect** the data using symbols.*

$v = 7{\cdot}2\ \text{m s}^{-1}$

$t = 4{\cdot}0\ \text{s}$ — ***Check** that the units are correct SI units.*

$a = ?$ — *Now look at the **symbols** – **which equation** includes these symbols: u, v, t and a?*

$v = u + at$ — ***Answer:** The first equation of motion.*

$\Rightarrow 7{\cdot}2 = 6{\cdot}4 + a \times 4{\cdot}0$ — ***Substitute** carefully.*

$\Rightarrow a = 0{\cdot}20\ \text{m s}^{-2}$ — *Do the **arithmetic**.*

***Check:** Correct **unit** and the right number of **significant figures** in the final answer.*

2 $u = 0\ \text{m s}^{-1}$ — ***Collect** data and check units.*

$t = 6{\cdot}0\ \text{s}$

$s = ?$ — *The value of g is given in the data page at the front of the Higher Physics paper.*

$a = g = 9{\cdot}8\ \text{m s}^{-2}$ — ***Which equation** has u, t, s and a?*

$s = ut + \tfrac{1}{2}at^2$

$\Rightarrow s = (0 \times 6{\cdot}0) + (\tfrac{1}{2} \times 9{\cdot}8 \times 6{\cdot}0^2)$ — ***Substitute** carefully.*

$= 0 + 176{\cdot}4$ — *Do the **arithmetic.***

$\Rightarrow s = 180\ \text{m}$ — ***Remember: Significant figures** and units!!*

3 $u = 0\ \text{m s}^{-1}$ — ***Collect** data and check units*

$v = 0{\cdot}0180\ \text{km s}^{-1} = 18{\cdot}0\ \text{m s}^{-1}$

$s = 0{\cdot}25\ \text{km} = 250\ \text{m}$

$a = ?$

$v^2 = u^2 + 2as$ — ***Which equation** has u, v, s and a?*

$\Rightarrow 18{\cdot}0^2 = 0^2 + (2 \times a \times 250)$ — ***Substitute** – be careful with squared terms!!*

$\Rightarrow a = 0{\cdot}648$ — ***Arithmetic.***

$a = 0{\cdot}648\ \text{m s}^{-2}$ — ***Significant figures** and **units**!!*

4 (a) $s = ut + \tfrac{1}{2}at^2$

$= ut + \tfrac{1}{2}\left(\dfrac{v-u}{t}\right)t^2$

$= ut + \tfrac{1}{2}vt - \tfrac{1}{2}ut$

$= \tfrac{1}{2}ut + \tfrac{1}{2}vt$

$= \tfrac{1}{2}(u + v)t$

(b) $s = \bar{v}t$ and $s = \tfrac{1}{2}(u + v)t$

Comparing these two relationships $\Rightarrow \bar{v} = \tfrac{1}{2}(u + v)$.

Exercises

Exercise 9 Projectiles

1 Initial velocity of ball, $u = 21\ \text{m s}^{-1}$.

Angle $\theta = 69°$ to the horizontal.

Initial horizontal component of velocity of the ball $= u \cos \theta = 21 \cos 69° = 7{\cdot}53\ \text{m s}^{-1}$.

Initial vertical component of velocity of ball $= u \sin \theta = 21 \sin 69° = 19{\cdot}6\ \text{m s}^{-1}$.

(a) Use vertical motion to find maximum height, i.e. $s_{vertical}$ when $v_{vertical} = 0\ \text{m s}^{-1}$.

$u_{vertical} = 19{\cdot}6\ \text{m s}^{-1}$ Use equation $v^2 = u^2 + 2as$

$v_{vertical} = 0\ \text{m s}^{-1}$ $0^2 = 19{\cdot}6^2 + (2 \times -9{\cdot}8 \times s)$

$a = g = -9{\cdot}8\ \text{m s}^{-2}$ ⇒ maximum height, $s = +19{\cdot}6 = 20\ \text{m}$

$s = ?$

(b) Use vertical motion to find time of flight, i.e. time when $v_{vertical} = -u_{vertical}$.

$u_{vertical} = 19{\cdot}6\ \text{m s}^{-1}$ Use equation $v = u + at$

$v_{vertical} = -19{\cdot}6\ \text{m s}^{-1}$ $-19{\cdot}6 = 19{\cdot}6 + (-9{\cdot}8 \times t)$

$a = g = -9{\cdot}8\ \text{m s}^{-2}$ ⇒ time of flight, $t = 4{\cdot}0\ \text{s}$

$t = ?$

(c) Now use horizontal motion: $v_{horizontal} = u_{horizontal} = 7{\cdot}53\ \text{m s}^{-1}$.

$t_{horizontal} = \text{time of flight} = 4{\cdot}0\ \text{s}$

$s_{horizontal} = vt = 7{\cdot}53 \times 4{\cdot}0 = 30{\cdot}1 = 30\ \text{m}$

Exercises

Exercise 10 Balanced and unbalanced forces

1 (a) Forces are balanced – uniform velocity in a straight line

(b) Forces are unbalanced – direction is changing.

(c) Forces are unbalanced – velocity is increasing.

(d) Forces are unbalanced – velocity is decreasing.

(e) Forces are balanced – electron is stationary.

(f) Forces are unbalanced – direction of electron is changing.

Exercises

Exercise 11 Newton's second law

1 $m = 1200$ kg $\qquad F = ma$
$a = 0{\cdot}70$ m s^{-2} $\qquad = 1200 \times 0{\cdot}70$
$F = ?$ $\qquad$ unbalanced force = 840 N

2 $a = 500$ mm s^{-2} = $0{\cdot}5$ mm s^{-2} $\qquad F = ma$ $\qquad$ *Remember to check units!!*
$F = 2{\cdot}0$ N $\quad 2{\cdot}0$ $\qquad = m \times 0{\cdot}5$
$m = ?$ $\qquad$ mass of trolley, $m = 4{\cdot}0$ kg

3 First find a: $u = 2{\cdot}4$ m s^{-1} $\qquad v = u + at$
$t = 5{\cdot}0$ s $\qquad \Rightarrow 6{\cdot}4 = 2{\cdot}4 + a \times 5{\cdot}0$
$v = 6{\cdot}4$ m s^{-1} $\qquad \Rightarrow a = 0{\cdot}80$ m s^{-2}
$a = ?$

Now use F $\qquad = ma$.
$m = 80$ kg $\qquad = 80 \times 0{\cdot}80$
unbalanced force = 64 N

Exercises

Exercise 12 Analysing forces

1 (a) *Buoyancy on its own is incorrect.*

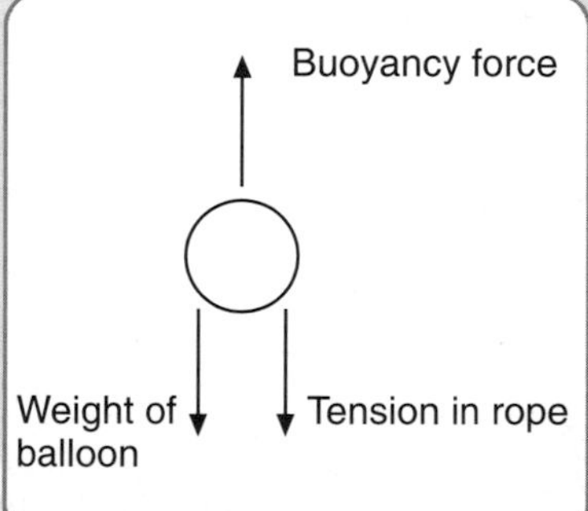

(b) The balloon is stationary so the forces are balanced
$\Rightarrow$ weight + tension = – buoyancy force.

Negative sign because the vectors act in opposite directions.

***Exercises** continued* ➢

Exercises continued

2 (a) $m = 1600$ kg $\quad F = ma$

$a = 1{\cdot}50 \text{ m s}^{-2} \quad = 1600 \times 1{\cdot}50$

$F = ? \quad = 2400$ N

(b) $m = 800$ kg $\quad F = ma$

$a = 1{\cdot}50 \text{ m s}^{-2} \quad = 800 \times 1{\cdot}50$

$F = ? \quad = 1200$ N

(c) (i) First sketch the forces acting on the caravan.

Unbalanced force = 1200 N

Tension in coupling = ?

Frictional forces = 300 N

unbalanced force = tension – frictional forces

$\Rightarrow 1200 = T - 300 \Rightarrow$ tension in coupling = 1500 N

(ii) Now sketch the forces acting on the car.

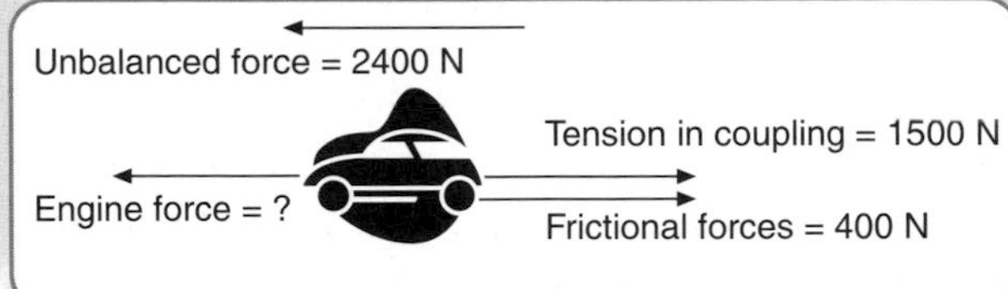

unbalanced force = engine force – tension – frictional forces

$\Rightarrow 2400 = F_E - 1500 - 400 \Rightarrow F_E = 4300$ N

Exercises

Exercise 13 Energy and power

1 (a) $m = 12{\cdot}0$ kg $\quad E_p = mgh$

$h = 15{\cdot}2 \text{ m} \quad = 12{\cdot}0 \times 9{\cdot}8 \times 15{\cdot}2$

$g = 9{\cdot}8 \text{ m s}^{-2} \quad = 1788$

$= 1800$ J

Note: The final answer is rounded to two figures as the value of g has only two significant figures.

Exercises continued ➢

Exercises continued

(b) Assuming no energy is lost/wasted the kinetic energy of the sphere just before it hits the ground is equal to its potential energy at the start $\Rightarrow E_k = 1800$ J.

(c) $E_k = 1800$ J $\quad E_k = \frac{1}{2}mv^2$

$m = 12{\cdot}0$ kg $\quad \Rightarrow \quad 1800 = \frac{1}{2} \times 12{\cdot}0 \times v^2$

$\Rightarrow \quad v = 17{\cdot}3 = 17$ m s^{-1}

(d) The kinetic energy of the sphere is converted to other forms of energy, such as heat energy in the sphere itself, in the surface of the ground where it lands and in the surrounding air. (Some sound is also produced but this is a tiny amount of energy.)

2 In the pendulum the energy of the swinging sphere is continually changing from kinetic energy to potential energy and back. To find maximum height assume the kinetic energy at the lowest point is equal to the potential energy at the highest point.

$\Rightarrow \quad mgh = \frac{1}{2}mv^2$ *Cancel m on both sides and substitute.*

$v = 900$ mm s^{-1} $\quad \Rightarrow \quad 9{\cdot}8 \times h = \frac{1}{2} \times 0{\cdot}900^2$

$= 0{\cdot}900$ m s^{-1} $\quad \Rightarrow \quad h = 0{\cdot}0413$ m $= 41$ mm

3 (a) $F = 1{\cdot}20$ kN $= 1200$ N $\quad E_w = Fd$

$d = 12{\cdot}9$ m $\quad = 1200 \times 12{\cdot}9$

$= 15\,480 = 15\,500$ J

(b) Use the sine of the angle to find the vertical distance the crate moves.

The final height of the crate above its starting point = $12{\cdot}9 \sin 18° = 3{\cdot}986$ m.

$g = 9{\cdot}8$ m s^{-2} $\quad E_p = mgh$

$m = 84{\cdot}0$ kg $\quad = 84{\cdot}0 \times 9{\cdot}8 \times 3{\cdot}986$

$= 3281 = 3300$ J

(c) energy wasted = work done by force – potential energy gained by crate

= 15 500 – 3300 = 12 200 J

(d) energy lost/wasted = $F_{friction} \times d$ *The energy is wasted overcoming $F_{friction}$.*

$\Rightarrow 12\,200 = F_{friction} \times 12{\cdot}9$

$\Rightarrow F_{friction} = 945{\cdot}7 = 950$ N

(e) (i) Input power is the power of the 1·20 kN force.

$t = 80$ s $\quad P = \dfrac{E}{t}$

$E_w = 15\,500$ J $\quad \Rightarrow P = \dfrac{15\,500}{80} = 193{\cdot}8 = 190$ W

(ii) Useful output power is the rate at which the crate gains potential energy.

$t = 80$ s $\quad P = \dfrac{E}{t}$

$E_p = 3300$ J $\quad \Rightarrow P = \dfrac{3300}{80} = 41{\cdot}25 = 41$ W

Exercises

Exercise 14 Momentum

1 (a)

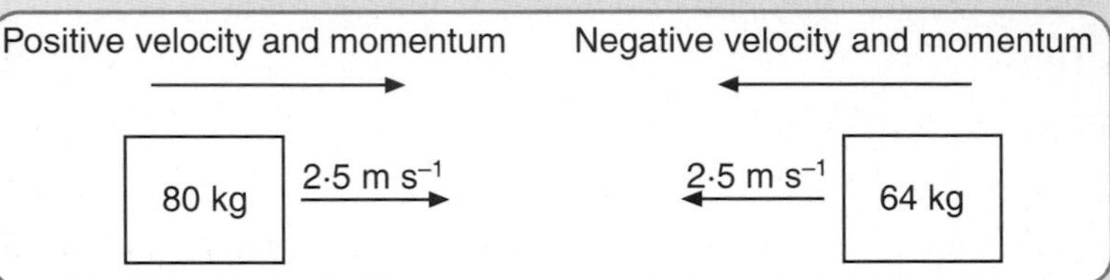

momentum of the male skater $= mv = 80 \times 2{\cdot}5 = +\ 200\ \text{kg m s}^{-1}$

momentum of the female skater $= mv = 64 \times -\ 2{\cdot}5 = -160\ \text{kg m s}^{-1}$

$\Rightarrow$ total momentum $= 200 - 160 = 40\ \text{kg m s}^{-1}$

(b) The total momentum is in the same direction as the motion of the male skater.

2 An unbalanced external force has caused the total momentum of the skaters to change. This could be due to one or both of the skaters pushing against the ice.

3

	Before collision		After collision	
Mass:	4·0 kg	8·0 kg	4·0 kg	8·0 kg
Velocity:	3·0 m s⁻¹	0 m s⁻¹	?	1·2 m s⁻¹
Momentum:	12·0 kg m s⁻¹	0 kg m s⁻¹	?	9·6 kg m s⁻¹

(a) total momentum before collision $= 12{\cdot}0\ \text{kg m s}^{-1}$

$=$ total momentum after collision

$\Rightarrow$ momentum of 4·0 kg trolley after collision $= (12{\cdot}0 - 9{\cdot}6) = 2{\cdot}4\ \text{kg m s}^{-1}$

$\Rightarrow$ velocity of 4·0 kg trolley after collision $= 2{\cdot}4/4{\cdot}0 = 0{\cdot}60\ \text{m s}^{-1}$

(b) Collision is inelastic as kinetic energy is not conserved.

total kinetic energy before collision $= (\frac{1}{2} \times 4{\cdot}0 \times 3{\cdot}0^2) + (\frac{1}{2} \times 8{\cdot}0 \times 0^2) = 18\ \text{J}$

total kinetic energy after collision $= (\frac{1}{2} \times 4{\cdot}0 \times 0{\cdot}6^2) + (\frac{1}{2} \times 8{\cdot}0 \times 1{\cdot}2^2)$
$= 6{\cdot}48\ \text{J}\ (= 6{\cdot}5\ \text{J})$

4 (a) Before collision: momentum of $m_1 = m_1u_1$ and momentum of $m_2 = m_2u_2$.

After collision: momentum of $m_1 = m_1v_1$ and momentum of $m_2 = m_2v_2$.

total momentum after collision = total momentum before collision

$\Rightarrow m_1v_1 + m_2v_2 = m_1u_1 + m_2u_2$

$\Rightarrow m_1v_1 - m_1u_1 = -\ m_2v_2 + m_2u_2$ *m_1 on one side, m_2 on the other.*

$\Rightarrow (m_1v_1 - m_1u_1) = -\ (m_2v_2 - m_2u_2)$ ***Equation** ①*

$\Rightarrow$ change in momentum of $m_1 = -$ change in momentum of m_2.

***Exercises** continued* ➢

Exercises continued

(b) From Newton's second law, $F = ma = m\left(\frac{v-u}{t}\right) = \frac{mv - mu}{t}$

$\Rightarrow Ft = mv - mu.$ **Equation** ②

Let the force on m_1 be F_1 for time t_1 and the force on m_2 be F_2 for time t_2.

$\Rightarrow F_1t_1 = m_1v_1 - m_1u_1$ and $F_2t_2 = m_2v_2 - m_2u_2$ *using* **Equation** ②

$\Rightarrow F_1t_1 = -F_2t_2$ *substituting in* **Equation** ①

$t_1 = t_2$ as both are the time for which m_1 and m_2 are in contact.

$\Rightarrow F_1 = -F_2$ *Cancel* t_1 *and* t_2.

i.e. the force exerted on m_1 is equal and opposite to the force exerted on m_2.

(c) The solutions to (a) and (b) assume that no external forces are acting.

Exercises

Exercise 15 Impulse

1 (a) $m = 48{\cdot}0\ \text{g} = 4{\cdot}80 \times 10^{-2}$ kg $\quad \Delta p = mv - mu$

$u = 0\ \text{m s}^{-1}$ $\quad = (4{\cdot}80 \times 10^{-2} \times 0{\cdot}380) - (4{\cdot}80 \times 10^{-2} \times 0)$

$v = 0{\cdot}380\ \text{m s}^{-1}$ $\quad = 1{\cdot}824 \times 10^{-2}$

$= 1{\cdot}82 \times 10^{-2}\ \text{kg m s}^{-1}$

(b) impulse on ball $= \Delta p = 1{\cdot}82 \times 10^{-2}$ N s

(c) $t = 6{\cdot}0\ \text{ms} = 6{\cdot}0 \times 10^{-3}$ s $\quad Ft = mv - mu$

$F = ?$ $\quad \Rightarrow F \times 6{\cdot}0 \times 10^{-3} = 1{\cdot}82 \times 10^{-2}$

$\Rightarrow F = 3{\cdot}03\ \text{N} = 3{\cdot}0\ \text{N}$ *Fewer significant figures here!*

2 (a) impulse = area under F–t graph

$= ½ \times 500 \times 12 \times 10^{-3}$ *The unit on the time axis is ms.*

$= 3{\cdot}0$ N s

(b) change in momentum of ball $= 3{\cdot}0\ \text{kg m s}^{-1}$ *i.e. equal to impulse*

(c) average force acting on ball $= \frac{\text{impulse}}{\text{total time}} = \frac{3{\cdot}0}{12 \times 10^{-3}} = 250$ N

Exercises

Exercise 16 Density

1 (a) *length* of cube side = 30·0 cm = 0·300 m $\Rightarrow V = 0{\cdot}300 \times 0{\cdot}300 \times 0{\cdot}300$
$= 0{\cdot}027\ 00\ \text{m}^3$

$\rho_{aluminium} = 2{\cdot}70 \times 10^3\ \text{kg m}^{-3}$ $\quad \rho = \dfrac{m}{V}$

$\Rightarrow \quad 2{\cdot}70 \times 10^3 = \dfrac{m}{0.027\ 00}$

$\Rightarrow \quad m = 72{\cdot}9\ \text{kg}$

(b) $m_{sea\ water} = 72{\cdot}9\ \text{kg}$ $\quad \rho = \dfrac{m}{V}$

$\rho_{sea\ water} = 1{\cdot}02 \times 10^3\ \text{kg m}^{-3}$ $\quad \Rightarrow \quad 1{\cdot}02 \times 10^3 = \dfrac{72.9}{V}$

$\Rightarrow \quad V = 0{\cdot}071\ 47 = 0{\cdot}0715\ \text{m}^3$

(c) $m_{air} = 72{\cdot}9\ \text{kg}$ $\quad \rho = \dfrac{m}{V}$

$\rho_{air} = 1{\cdot}29\ \text{kg m}^{-3}$ $\quad \Rightarrow \quad 1.29 = \dfrac{72.9}{V}$

$\Rightarrow \quad V = 56{\cdot}51 = 56{\cdot}5\ \text{m}^3$

(d) The volume of air is much greater because the particles in air are much further apart than the particles in sea water.

2 (a) $m = 350\ \text{kg}$ $\quad \rho = \dfrac{m}{V}$

$\rho_{ice} = 9{\cdot}20 \times 10^2\ \text{kg m}^{-3}$ $\quad \Rightarrow \quad 9{\cdot}20 \times 10^2 = \dfrac{350}{V}$

$\Rightarrow \quad V = 0{\cdot}380 = 0{\cdot}38\ \text{m}^3$

(b) The volume of the melt water is less than the volume of the ice in the sculpture.

The mass of the melt water is equal to the mass of the ice.

The density of water is greater than the density of ice

$\Rightarrow$ the mass is more tightly packed in water than in ice.

Exercises

Exercise 17 Measuring the density of air

1 (a) $V_{air} = 750\text{ ml} = 750 \times 10^{-6}\text{ m}^3$ (1 ml = 1 cm^3)

$= 7{\cdot}50 \times 10^{-4}\text{ m}^3$

$m_{empty\ syringe} = 430{\cdot}165\text{ g}$

$m_{syringe + air} = 431{\cdot}133\text{ g}$

$\Rightarrow m_{air} = 0{\cdot}968\text{ g} = 9{\cdot}68 \times 10^{-4}\text{ kg}$

$$\rho = \frac{m}{V}$$

$$= \frac{9.68 \times 10^{-4}}{7.50 \times 10^{-4}}$$

$$= 1{\cdot}291 = 1{\cdot}29\text{ kg m}^{-3}$$

(b) Pull the plunger further back to increase the volume of air inside the syringe. This increases both the volume and mass measurements.

The absolute uncertainties in the measurement of mass and volume do not change

$\Rightarrow$ percentage uncertainties in the measurements of mass and volume are less

$\Rightarrow$ percentage uncertainty in the calculated value of density is less.

Exercises

Exercise 18 Pressure

1 Total area of tyres in contact with ground, $A = (0{\cdot}032 \times 4) = 0{\cdot}128\text{ m}^2$.

$P = 9{\cdot}2 \times 10^4\text{ Pa}$

$$P = \frac{F}{A}$$

$$\Rightarrow 9{\cdot}2 \times 10^4 = \frac{F}{0.128}$$

$$\Rightarrow F = 1{\cdot}178 \times 10^4\text{ N} = mg \qquad F = \textit{weight of car}$$

$$\Rightarrow m = \frac{1{\cdot}178 \times 10^4}{9{\cdot}8} = 1{\cdot}202 \times 10^3 = 1200\text{ kg}$$

2 $1\text{ cm} = 0{\cdot}01\text{ m} \Rightarrow 1\text{ cm}^2 = 0{\cdot}01 \times 0{\cdot}01\text{ m}^2 = 1{\cdot}0 \times 10^{-4}\text{ m}^2$

$A = 0{\cdot}64\text{ cm}^2 = 0{\cdot}64 \times 1{\cdot}0 \times 10^{-4}\text{ m}^2 = 6{\cdot}4 \times 10^{-5}\text{ m}^2$

$m = 46\text{ kg}$

$g = 9{\cdot}8\text{ m s}^{-2}$

$$P = \frac{F}{A}$$

$$= \frac{46 \times 9{\cdot}8}{6{\cdot}4 \times 10^{-5}}$$

$$= 7{\cdot}043 \times 10^6 = 7{\cdot}0 \times 10^6\text{ Pa}$$

Exercises

Exercise 19 Pressure in fluids

1 (a) $\rho = 1{\cdot}00 \times 10^3\ \text{kg m}^{-3}$ $\quad P = g\rho h + \text{air pressure}$

$h = 25\ \text{m}$ $\quad = (9{\cdot}8 \times 1{\cdot}00 \times 10^3 \times 25) + 1{\cdot}01 \times 10^5$

$g = 9{\cdot}8\ \text{m s}^{-2}$ $\quad = 3{\cdot}46 \times 10^5 = 3{\cdot}5 \times 10^5\ \text{Pa}$

(b) (i) $A = (0{\cdot}30 \times 0{\cdot}40)\ \text{m}^2$ $\quad P = \dfrac{F}{A}$

$\Rightarrow \quad 3{\cdot}5 \times 10^5 = \dfrac{F}{0.12}$

$\Rightarrow \quad F = 4{\cdot}2 \times 10^4\ \text{N}$

(ii) The solution assumes that the pressure at the centre of the window is equal to the average pressure over the whole area of the window.

2 (a) The readings on the balance are less than the weight of the metal block. This is due to the upward buoyancy force acting on the metal block.

(b) The reading on the balance is less when the metal block is suspended in sea water.

(c) The upthrust depends on the difference in pressure between the upper and lower faces of the block. The dimensions of the block do not change and the block is suspended at the same depth in each liquid. Sea water has a higher density than fresh water so the difference in pressure is greater for sea water.

(d) There is no difference in the balance readings compared with the original experiment. The difference in depth between the upper and lower sides of the block is the same in both experiments (this assumes that there is no change in the orientation of the metal block between the experiments) ⇒ the difference in pressure between the upper and lower sides is the same ⇒ the upward buoyancy force is the same in both experiments.

3 Air pressure at sea level is due to the weight of the air above the surface of the sea. There is less air above the top of a mountain ⇒ the weight of the air is less ⇒ P is less.

Exercises

Exercise 20 Kelvin temperature scale

1 (a) 373 K (b) 173 K (c) 546 K (d) 3 K (e) 300 K

2 (a) – 173 °C (b) 127 °C (c) – 3 °C (d) – 200 °C (e) 77 °C

3 (a) change in temperature = (350 – 20) = 330 °C

(b) change in temperature = 330 K

Exercises

Exercise 21 Gas laws

1 (a) $V_1 = 3V_2$ $\qquad \dfrac{P_1V_1}{T_1} = \dfrac{P_2V_2}{T_2}$

$P_1 = 8{\cdot}17 \times 10^4$ Pa $\qquad \Rightarrow \dfrac{8{\cdot}17 \times 10^4 \times 3V_2}{T_1} = \dfrac{P_2 \times V_2}{T_1}$

$T_1 = T_2$ $\qquad \Rightarrow P_2 = 2{\cdot}451 \times 10^5 = 2{\cdot}45 \times 10^5$ Pa

(b) The final pressure of the gas is higher than the value calculated in part (a). The pressure of a fixed mass of gas is directly proportional to temperature when volume is constant ⇒ the increase in temperature causes the pressure to rise more.

2 (a) $h = 32$ m $\qquad P = g\rho h$ + air pressure

$g = 9{\cdot}8$ m s^{-2} $\qquad = (9{\cdot}8 \times 1{\cdot}02 \times 10^3 \times 32) + 1{\cdot}01 \times 10^5$

$\rho = 1{\cdot}02 \times 10^3$ kg m^{-3} $\qquad = 4{\cdot}208 \times 10^5 = 4{\cdot}2 \times 10^5$ Pa

(b) $V_1 = 20$ m^3 $\qquad$ *bell is initially full of air*

$P_1 = 1{\cdot}01 \times 10^5$ Pa $\qquad \dfrac{P_1V_1}{T_1} = \dfrac{P_2V_2}{T_2}$

$$\Rightarrow \frac{1.01 \times 10^5 \times 20}{T_1} = \frac{4.2 \times 10^5 \times V_2}{T_1}$$

$P_2 = 4{\cdot}2 \times 10^5$ Pa $\qquad V_2 = 4{\cdot}809 = 4{\cdot}8$ m^3

$T_1 = T_2$

3 The relationship can be established in two ways – by drawing a graph or by using mathematics. Both methods are shown below.

By mathematics: First convert all of the temperatures to kelvins; then find the ratio $\frac{P}{T}$ for each pair of pressure and temperature readings. It is easiest to do this by copying the table and then adding extra rows.

Temperature/°C	20	30	40	50	60	70
Temperature/K	293	303	313	323	333	343
Pressure/× 10⁵ Pa	1·010	1·044	1·079	1·113	1·148	1·182
$\frac{P}{T}$ × 10² Pa K⁻¹	3·447	3·446	3·447	3·446	3·447	3·446

The ratio $\frac{P}{T}$ is constant ⇒ pressure is directly proportional to kelvin temperature.

Exercises *continued* ➢

Exercises continued

By drawing graph: Plot all of the points on a graph with *P* on the *y*-axis. Graph 1 is obtained. This is a straight line graph but it does not go through the origin.

Extend the graph backwards until it cuts the *x*-axis.

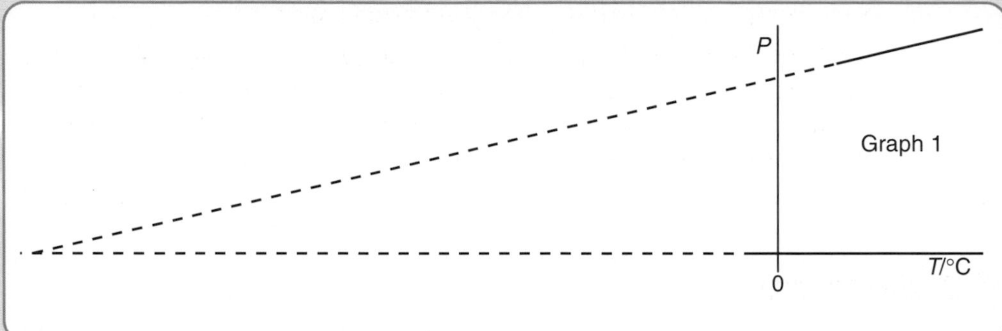

Now move the pressure axis so that the zero of temperature is at this point. Graph 2 is obtained.

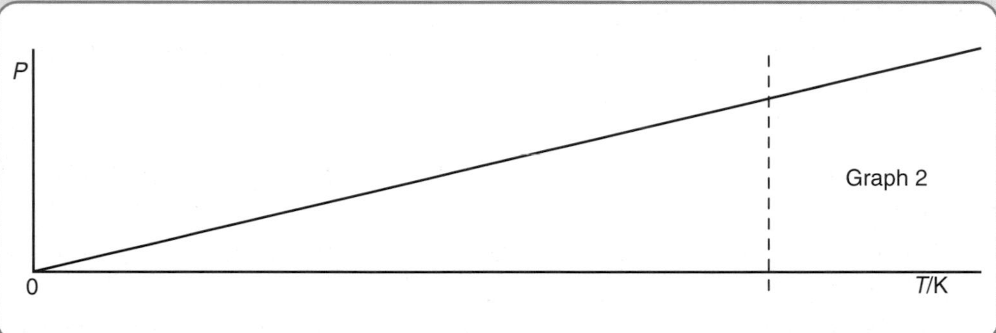

Graph 2 is a straight line graph that passes through the origin. Thus the pressure of the gas is directly proportional to the temperature provided that it is measured from the new zero. Notice that in graph 2 the unit on the temperature axis is the kelvin.

Exercises

Exercise 22 Kinetic theory

1 Ice and water are made up of the same particles. The density of ice is less than the density of water as the particles are slightly further apart in the ice than they are in the water.

2 During the journey the temperature of the air inside the tyre rises. The average kinetic energy of the particles increases and the particles get faster. The particles hit the walls of the tyre with greater force and more often each second. This causes the pressure inside the tyre to increase.

Exercises continued ➢

Exercises continued

3 (a) There are two effects as the balloon gets higher: the air pressure decreases and the temperature drops. The volume of the balloon increases until the pressure inside the balloon is the same as the pressure outside the balloon. Part of the decrease in pressure inside the balloon is due to the decrease in temperature.

(b) As the temperature falls, the average kinetic energy of the particles decreases and the particles get slower. Also, as the volume of the balloon increases the spacing of the particles increases. The particles inside the balloon hit the sides of the balloon with less force and less often each second.

5.2 Electricity

Exercises

Exercise 23 Electric fields

1 The potential difference between the terminals of the supply = 15 V. (When $Q = 1$ C, $V = W$.)

2 $Q = 40$ mC $= 4{\cdot}0 \times 10^{-2}$ C $W = QV$
$V = 1{\cdot}8$ kV $= 1{\cdot}8 \times 10^3$ V $= 4{\cdot}0 \times 10^{-2} \times 1{\cdot}8 \times 10^3$ *Careful with the units!*
$W = ?$ $= 72$ J

3 (a) $Q = 0{\cdot}30$ C $W = QV$
$W = 1{\cdot}2$ J $\Rightarrow$ $1{\cdot}2 = 0{\cdot}30 \times V$
$V = ?$ $\Rightarrow$ $V = 4{\cdot}0$ V

(b) The work done in moving the 0·30 C from A to D and then C = 1·2 J.

The work done in moving charge in an electric field depends only on the starting and finishing points.

4 (a) Kinetic energy gained by electrons = work done, $E_k = W = QV$.
$Q = 1{\cdot}60 \times 10^{-19}$ C $\Rightarrow$ $E_k = 1{\cdot}60 \times 10^{-19} \times 2{\cdot}30 \times 10^3$
$V = 2{\cdot}30$ kV $= 2{\cdot}30 \times 10^3$ V $= 3{\cdot}68 \times 10^{-16}$ J

(b) $m_e = 9{\cdot}11 \times 10^{-31}$ kg $E_k = \frac{1}{2}mv^2$
$\Rightarrow$ $3{\cdot}68 \times 10^{-16} = \frac{1}{2} \times 9{\cdot}11 \times 10^{-31} \times v^2$
$\Rightarrow$ $v = 2{\cdot}842 \times 10^7 = 2{\cdot}84 \times 10^7$ m s^{-1}

Assumption: The electrons are initially at rest.

Exercises

Exercise 24 e.m.f. and internal resistance

1 (a) $E = 20$ V $\quad E = I(R + r)$

$r = 2{\cdot}0\ \Omega \quad \Rightarrow \quad 20 = I(38 + 2)$

$R = 38\ \Omega \quad \Rightarrow \quad I = 0{\cdot}50$ A

(b) $P = I^2R$

$= 0{\cdot}50^2 \times 38$

$= 9{\cdot}5$ W

(c) Power is wasted in the internal resistance $P = I^2r$

$= 0{\cdot}50^2 \times 2{\cdot}0$

$= 0{\cdot}50$ W.

2 $E = 12$ V $\quad E = IR + Ir$

$I = 0{\cdot}50$ A $\quad \Rightarrow \quad 12 = (0{\cdot}50 \times 23) + (0{\cdot}50 \times r)$

$R = 23\ \Omega \quad \Rightarrow \quad r = 1{\cdot}0\ \Omega$

3 (a) The student is correct. The potential difference across the terminals of a supply $V = E - Ir$.

When $I = 0$, $Ir = 0$ and $V = E$. For any other value of I, $V < E$.

(b) $V = E - Ir$. When the current is zero, terminal p.d. = e.m.f.

As I increases $\Rightarrow$ Ir increases $\Rightarrow$ terminal p.d. decreases.

Exercises

Exercise 25 Measuring e.m.f. and internal resistance

1 Plot a graph of terminal p.d. against current.

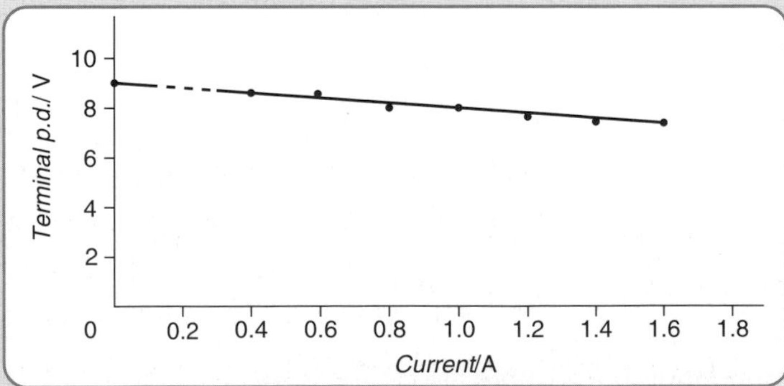

(a) Extend the graph to the y-axis $\Rightarrow$ when $I = 0$ the terminal p.d. = 9·0 V

$\Rightarrow$ e.m.f. = 9·0 V.

(b) gradient $= \dfrac{y_2 - y_1}{x_2 - x_1} = \dfrac{7{\cdot}40 - 8{\cdot}60}{1{\cdot}60 - 0{\cdot}40} = \dfrac{-1{\cdot}2}{1{\cdot}2} = -1{\cdot}0 \Rightarrow$ internal resistance = 1·0 Ω

Note: The two points used to find the gradient must be on the line of the graph.

Exercises

Exercise 26 Series and parallel circuits

1 The p.d. across the terminals of the supply = 20 V.

2 (a) $R_1 = 4{\cdot}0\ \Omega$ $\quad R = R_1 + R_2 + R_3$

$R_2 = 6{\cdot}0\ \Omega$ $\quad = (4{\cdot}0 + 6{\cdot}0 + 10) = 20\ \Omega$

$R_3 = 10\ \Omega$ $\quad V = IR$

$V = 12\ \text{V}$ $\quad \Rightarrow \quad 12 = I \times 20$

$\Rightarrow$ current drawn from battery $I = 0{\cdot}60$ A

(b) p.d. across $R_1 = IR_1 = 0{\cdot}60 \times 4{\cdot}0 = 2{\cdot}4$ V

p.d. across $R_2 = IR_2 = 0{\cdot}60 \times 6{\cdot}0 = 3{\cdot}6$ V

p.d. across $R_3 = IR_3 = 0{\cdot}60 \times 10 = 6{\cdot}0$ V

(c) Sum of p.d.s across resistors = 2·4 + 3·6 + 6·0 V = 12 V.

The sum of the p.d.s across the resistors is equal to the p.d. across the battery terminals.

3 (a) $R_1 = 4{\cdot}0\ \Omega$ $\quad \dfrac{1}{R} = \dfrac{1}{R_1} + \dfrac{1}{R_2} + \dfrac{1}{R_3}$

$R_2 = 6{\cdot}0\ \Omega$ $\quad \Rightarrow \quad = \left(\dfrac{1}{4{\cdot}0} + \dfrac{1}{6{\cdot}0} + \dfrac{1}{10}\right) = 0{\cdot}517$

$\Rightarrow \quad R = 1{\cdot}93\ \Omega$ *Remember to invert!!*

$R_3 = 10\ \Omega$ $\quad V = IR$

$V = 12\ \text{V}$ $\quad \Rightarrow \quad 12 = I \times 1{\cdot}93$

$\Rightarrow$ current drawn from battery $I = 6{\cdot}22 = 6{\cdot}2$ A

(b) current in $R_1 = \dfrac{V}{R_1} = \dfrac{12}{4{\cdot}0} = 3{\cdot}0$ A

current in $R_2 = \dfrac{V}{R_2} = \dfrac{12}{6{\cdot}0} = 2{\cdot}0$ A

current in $R_3 = \dfrac{V}{R_3} = \dfrac{12}{10} = 1{\cdot}2$ A

(c) Sum of currents in resistors = (3·0 + 2·0 + 1·2) = 6·2 A.

The sum of the currents in the resistors is equal to the current drawn from the battery.

4 (a) The p.d. across the parallel branches is equal to the p.d. across the 4·0 Ω resistor.

$R_1 = 4{\cdot}0\ \Omega$ $\quad V = IR$

$I_1 = 0{\cdot}30\ \text{A}$ $\quad = 0{\cdot}30 \times 4{\cdot}0 = 1{\cdot}2$ V

Exercises *continued* ➢

Exercises continued

(b) The p.d. across the 6·0 Ω resistor is equal to the p.d. across the 4·0 Ω resistor.

$R_2 = 6{\cdot}0\ \Omega \quad \Rightarrow \quad 1{\cdot}2 = I_2 \times 6{\cdot}0$

$I_2 = 0{\cdot}20$ A

(c) Current in 10 Ω resistor is equal to the sum of the currents in the parallel branches.

$I = I_1 + I_2$

$= 0{\cdot}30 + 0{\cdot}20$

$= 0{\cdot}50$ A

(d) p.d. across battery = p.d. across the 10 Ω resistor + p.d. across the parallel branches

$= (0{\cdot}50 \times 10) + 1{\cdot}2$

$= 6{\cdot}2$ V

5 (a) The 6·0 kΩ resistor is in series with the 4·0 kΩ resistor ⇒ current is the same in both.

Total resistance of the 4·0 kΩ/6·0 kΩ branch = 10·0 kΩ.

p.d. across the battery = p.d. across the 4·0 kΩ/6·0 kΩ branch

$I = 3{\cdot}0 \times 10^{-4}$ A $\quad V = IR$

$R = 1{\cdot}0 \times 10^{4}\ \Omega \quad = 3{\cdot}0 \times 10^{-4} \times 1{\cdot}0 \times 10^{4}$

$= 3{\cdot}0$ V

(b) First calculate the p.d. across the 6·0 kΩ resistor – this is the p.d. across AX.

$R = 6{\cdot}0 \times 10^{3}\ \Omega \quad V_{AX} = IR$

$= 3{\cdot}0 \times 10^{-4} \times 6{\cdot}0 \times 10^{3} = 1{\cdot}8$ V

Now use battery voltage to calculate the current in the 5·0 kΩ/5·0 kΩ branch.

Total resistance of the 5·0 kΩ/5·0 kΩ branch = 10·0 kΩ.

$R = 1{\cdot}0 \times 10^{4}\ \Omega \quad V = IR$

$\Rightarrow \quad 3{\cdot}0 = I \times 1{\cdot}0 \times 10^{4}$

$\Rightarrow \quad I = 3{\cdot}0 \times 10^{-4}$ A

Now calculate the p.d. across the upper 5·0 kΩ resistor – this is the p.d. across AY.

$V_{AY} = IR$

$= 3{\cdot}0 \times 10^{-4} \times 5{\cdot}0 \times 10^{3} = 1{\cdot}5$ V

⇒ p.d. across XY $= V_{AX} - V_{AY}$

$= 1{\cdot}8 - 1{\cdot}5 = 0{\cdot}3$ V

Exercises

Exercise 27 Wheatstone bridge

1 $R_1 = 20\ \Omega$ $\qquad \frac{R_1}{R_2} = \frac{R_3}{R_4}$

$R_2 = 60\ \Omega$ $\qquad \Rightarrow \frac{20}{60} = \frac{18}{R_4}$

$R_3 = 18\ \Omega$ $\qquad \Rightarrow R_4 = 54\ \Omega$

2 (a) Consider the branch with the 12 kΩ and 18 kΩ resistors.

Total resistance = $R_1 + R_2 = (12 + 18)\ \text{k}\Omega = 30\ \text{k}\Omega$. *The resistors are in series.*

Using $V = IR \Rightarrow$ current in this branch = 5×10^{-4} A.

Using $V = IR$ again $\Rightarrow$ p.d. across the 12 kΩ resistor = 6·0 V.

Assumption: The current in the voltmeter is negligible.

(b) The p.d. across the 18 kΩ resistor = 15 – 6·0 = 9·0 V. *i.e.* $V_{supply} - V_{12\ \text{k}\Omega}$

(c) Consider the branch with the 2700 Ω and 5400 Ω resistors.

$$\text{p.d. across the 2700 } \Omega \text{ resistor} = V_{supply} \times \frac{R_1}{R_1 + R_2} = 15 \times \frac{2700}{(2700 + 5400)} = 5{\cdot}0\ \text{V}$$

Note: This method uses one of the potential divider relationships. It is quicker than the method used in the solution to part (a). This method is OK because the current in the high resistance voltmeter is negligible. If you want to check, use the method used in part (a) to confirm that the answer is correct.

The voltmeter reading = $V_{12\ \text{k}\Omega} - V_{2700\ \Omega} = (6{\cdot}0 - 5{\cdot}0) = 1{\cdot}0$ V.

(d) The voltmeter reading is zero when the bridge is balanced.

$R_1 = 12\ \text{k}\Omega = 12\,000\ \Omega$ $\qquad \frac{R_1}{R_2} = \frac{R_3}{R_4}$

$R_2 = 18\ \text{k}\Omega = 18\,000\ \Omega$ $\qquad \Rightarrow \frac{12\,000}{18\,000} = \frac{R_3}{5400}$

$R_4 = 5400\ \Omega$ $\qquad \Rightarrow R_3 = 3600\ \Omega$

(e) The voltmeter reading decreases from its initial values of 1 V.

When the resistance of the variable resistor is 3600 Ω the voltmeter reading is zero.

When the resistance of the variable resistor is greater than 3600 Ω the voltmeter reading is negative, i.e. the direction of the p.d. across XY has reversed.

The voltage at X is constant. As the resistance of the variable resistor is increased the voltage at Y decreases from a value greater than the voltage at X to a value lower than the voltage at X.

Exercises *continued* ➢

Exercises continued

3 (a) $R_1 = 3000\ \Omega$ $\qquad \dfrac{R_1}{R_2} = \dfrac{R_3}{R_4}$

$R_2 = 4500\ \Omega$ $\qquad \Rightarrow \dfrac{3000}{4500} = \dfrac{R_3}{3600}$

$R_4 = 3600\ \Omega$ $\qquad \Rightarrow R_3 = 2400\ \Omega$

(b) Assume the resistance of the thermistor falls as temperature rises.

The p.d.s across AX and BX do not change and the resistance of the variable resistor does not change.

At 10°C the resistance of the thermistor is greater than 3600 Ω. The p.d. across BY is greater than the p.d. across BX ⇒ point Y is at a higher voltage than point X.

As the temperature is increased the resistance of the thermistor falls and the p.d. between X and Y decreases.

At 15°C the p.d. between X and Y is zero.

When the temperature is above 15°C, the resistance of the thermistor is less than 3600 Ω.

The p.d. across BY is less than the p.d. across BX so that point Y is at a lower voltage than point X.

Exercises

Exercise 28 Alternating current and voltage

1 $V_{peak} = 140\ \text{V}$

$$V_{peak} = \sqrt{2}\, V_{r.m.s.}$$
$$\Rightarrow 140 = \sqrt{2}\, V_{r.m.s.}$$
$$\Rightarrow V_{r.m.s.} = 98{\cdot}99\ \text{V} = 99{\cdot}0\ \text{V}$$

2 (a) $I_{r.m.s.}$ in 20 Ω resistor = $I_{r.m.s.}$ drawn from the supply

Total resistance of circuit: $R = R_1 + R_2$

$$= 15 + 20 = 35\ \Omega$$
$$V_{r.m.s.} = I_{r.m.s.}R$$
$$\Rightarrow 7{\cdot}0 = I_{r.m.s.} \times 35$$
$$\Rightarrow I_{r.m.s.} = 0{\cdot}20\ \text{A}$$

(b) I_{peak} in the 15 Ω resistor = I_{peak} drawn from the supply

$$I_{peak} = \sqrt{2} I_{r.m.s.}$$
$$= \sqrt{2} \times 0{\cdot}20$$
$$= 0{\cdot}2828 = 0{\cdot}28\ \text{A}$$

Exercises continued ➢

Exercises continued

(c) Remember to use r.m.s. for the power calculation.

$$P = (I_{r.m.s.})^2 R$$
$$= 0{\cdot}20^2 \times 15 = 0{\cdot}60 \text{ W}$$

3 First calculate the total resistance of the circuit.

Parallel branches: $$\frac{1}{R} = \frac{1}{R_1} + \frac{1}{R_2}$$

$$= \frac{1}{12} + \frac{1}{6}$$

$$\Rightarrow R_{parallel} = 4{\cdot}0\ \Omega$$

Total external resistance of circuit: $$R = R_1 + R_{parallel}$$
$$= 3{\cdot}8 + 4{\cdot}0$$
$$= 7{\cdot}8\ \Omega$$

Now calculate r.m.s. e.m.f. $$E_{r.m.s.} = I_{r.m.s.}(R + r)$$
$$= 2{\cdot}5(7{\cdot}8 + 2{\cdot}2)$$
$$= 25 \text{ V}$$

Finally, calculate the peak e.m.f. $$E_{peak} = \sqrt{2}\, E_{r.m.s.}$$
$$= \sqrt{2} \times 25$$
$$= 35{\cdot}35 = 35 \text{ V}$$

4 (a) amplitude of oscilloscope trace = 2·5 divisions $\Rightarrow V_{peak} = 2{\cdot}5 \times 2 = 5{\cdot}0$ V

(b) $f = 200$ Hz $$T = \frac{1}{f} = \frac{1}{200} = 5{\cdot}0 \times 10^{-3} \text{ s} = 5{\cdot}0 \text{ ms}$$

There are five full waves on screen ⇒ time for trace to cross the screen = 5 × 5·0 = 25·0 ms.

There are ten divisions on screen ⇒ time-base setting = 2·5 ms per division.

(c) The trace now displays 12·5 waves. There is no change in the amplitude of the trace.

Note: The question does not ask for explanation, only description.

Exercises

Exercise 29 Capacitance

1 (a) $C = 40\ \mu F = 40 \times 10^{-6}\ F$ $\qquad C = \frac{Q}{V}$

$V = 20\ V \qquad \Rightarrow \qquad 40 \times 10^{-6} = \frac{Q}{20}$

$\Rightarrow \qquad Q = 8{\cdot}0 \times 10^{-4}\ C$

(b) $E = ½QV$

$= ½ \times 8{\cdot}0 \times 10^{-4} \times 20$

$= 8{\cdot}0 \times 10^{-3}\ J$

2 $C = 250\ nF = 250 \times 10^{-9}\ F$ $\qquad E = ½CV^2$

$V = 120\ V$ $\qquad = ½ \times 250 \times 10^{-9} \times 120^2$

$= 1{\cdot}8 \times 10^{-3}\ J$

3 The plates of a capacitor contain many millions of charges, i.e. protons and electrons. In neutral plates the total positive charge and total negative charge are equal. Excess charge on the plates is the difference between the total positive charge and the total negative charge.

4 $I = 200\ \mu A = 200 \times 10^{-6}\ A$ $\qquad Q = It$

$t = 20\ s$ $\qquad = 200 \times 10^{-6} \times 20$

$= 4{\cdot}0 \times 10^{-3}\ C$

$C = 0{\cdot}50\ mF = 5{\cdot}0 \times 10^{-4}\ F$ $\qquad C = \frac{Q}{V}$

$\Rightarrow \qquad 5{\cdot}0 \times 10^{-4} = \frac{4 \times 10^{-3}}{V}$

$\Rightarrow \qquad V = 8{\cdot}0\ V$

Exercises

Exercise 30 Capacitors in circuits

1 (a) The ammeter reading starts high and gradually falls to zero. The voltmeter reading starts high and gradually falls to zero.

(b)

(i)

(ii)

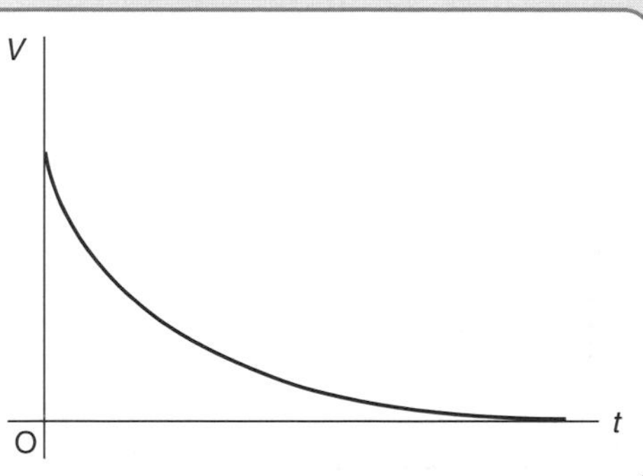

2 (a) Initial p.d. across the capacitor = 0 V.

(b) Initially all of the battery e.m.f. is across the resistor.

$V = 24$ V $\qquad$ $V = IR$

$R = 600\ \Omega$ $\quad \Rightarrow \quad 24 = I \times 600$

$\Rightarrow \quad I = 0{\cdot}040$ A

(c) First find p.d. across resistor. $\quad V = IR$

$= 0{\cdot}010 \times 600 = 6{\cdot}0$ V

$\Rightarrow$ p.d. across capacitor = 24 – 6·0 = 18 V $\qquad$ *$V_{battery}$ – p.d. across R.*

(d) Final p.d. across resistor = 0 V. $\qquad$ *p.d. across resistor = IR and final I = 0 A.*

(e) Final p.d. across capacitor = 24 V.

(f) $C = 200\ \mu F = 200 \times 10^{-6}$ F $\qquad E = ½CV^2$

$= ½ \times 200 \times 10^{-6} \times 24^2$

$= 0{\cdot}0576 = 0{\cdot}058$ J

3

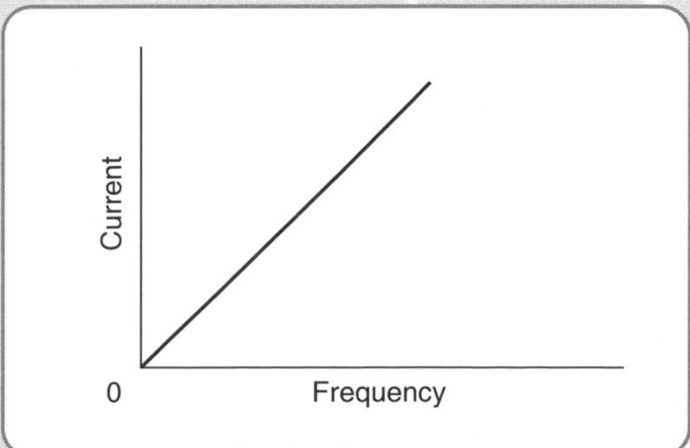

Exercises

Exercise 31 Op-amps in inverting mode

1 (a) $R_f = 40\ \text{k}\Omega$ $\qquad \dfrac{V_o}{V_1} = -\dfrac{R_f}{R_1}$

$R_1 = 5{\cdot}0\ \text{k}\Omega$ $\qquad \Rightarrow \dfrac{-12}{V_1} = -\dfrac{40}{5{\cdot}0}$

$V_o = -12\ \text{V}$ $\qquad \Rightarrow V_1 = 1{\cdot}5\ \text{V}$

(b) new $V_1 = 3{\cdot}0\ \text{V}$ $\qquad \dfrac{V_o}{V_1} = -\dfrac{R_f}{R_1}$

$$\Rightarrow \frac{V_o}{3{\cdot}0} = -\frac{40}{5{\cdot}0}$$

$$\Rightarrow V_o = -24\ \text{V}$$

The predicted value is less than the negative supply voltage

$\Rightarrow$ the amplifier is saturated $\Rightarrow V_o = -15$ V.

2 (a) First calculate the peak input voltage.

$V_{r.m.s.} = 2{\cdot}8\ \text{V}$ $\qquad V_{peak} = \sqrt{2}\, V_{r.m.s.}$

$$= \sqrt{2} \times 2{\cdot}8$$

$$= 3{\cdot}96\ \text{V}$$

$V_1 = 3{\cdot}96\ \text{V}$ $\qquad \dfrac{V_o}{V_1} = -\dfrac{R_f}{R_1}$

$R_f = 140\ \text{k}\Omega$ $\qquad \Rightarrow \dfrac{V_o}{3{\cdot}96} = -\dfrac{140}{40}$

$R_1 = 40\ \text{k}\Omega$ $\qquad \Rightarrow V_o = -13{\cdot}86\ \text{V}$ *This is not rounded as it is not a final answer.*

The predicted peak voltage is less than the negative supply voltage $\Rightarrow$ the amplifier is saturated $\Rightarrow$ peak output voltage = -12 V.

(b)

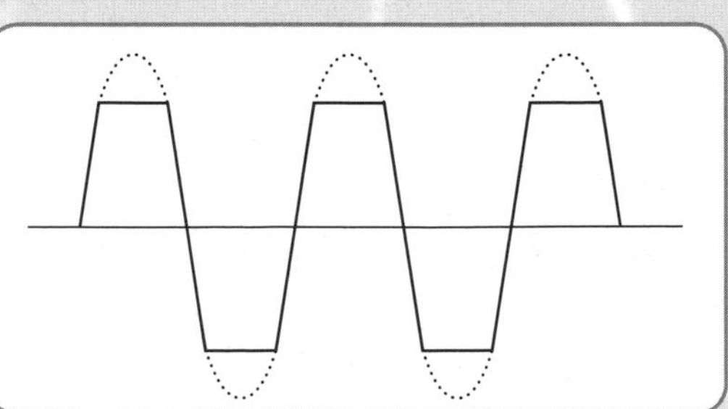

(c) The output is now the complete a.c. voltage wave.

When the supply voltage is changed to ± 16 V, predicted peak output voltage is now greater than the negative supply voltage (and less than the positive supply voltage).

Exercises

Exercise 32 Op-amps in differential mode

1 (a) First calculate the p.d. across the 3·0 kΩ resistor. $V_2 = V \times \frac{R_1}{R_1 + R_2}$

$$= 9{\cdot}0 \times \frac{3000}{3000 + 6000}$$

$$= 3{\cdot}0\ \text{V}$$

Now do the op-amp calculation.

$V_o = -0{\cdot}75$ V $\qquad V_o = (V_2 - V_1)\frac{R_f}{R_1}$

$R_f = 20$ kΩ $\qquad \Rightarrow \quad -0{\cdot}75 = (3{\cdot}0 - V_1) \times \frac{20\,000}{80\,000}$

$R_1 = 80$ kΩ $\qquad \Rightarrow \quad V_1 = 6{\cdot}0$ V

(b) p.d. across LDR $= V \times \frac{R_1}{R_1 + R_2}$ *Use potential divider equation.*

$$\Rightarrow \quad 6{\cdot}0 = 9{\cdot}0 \times \frac{R_{LDR}}{R_{LDR} + 2000}$$

$$\Rightarrow \quad 6{\cdot}0(R_{LDR} + 2000) = 9{\cdot}0R_{LDR}$$

$$\Rightarrow \quad 6{\cdot}0R_{LDR} + 12\,000 = 9{\cdot}0R_{LDR}$$

$$\Rightarrow \quad R_{LDR} = 4000\ \Omega$$

(c) When the intensity of light incident on the LDR is increased the resistance of the LDR decreases. This causes the p.d. across the LDR, V_1, to decrease.

This in turn causes V_o to increase (become less negative).

When V_1 becomes lower than V_2, i.e. the p.d. across the 3·0 kΩ resistor, the output voltage of the op-amp becomes positive.

2 (a) The fixed potential divider keeps V_2, the p.d. of the non-inverting input, constant.

As the temperature falls the resistance of the thermistor increases

⇒ the p.d. across the thermistor increases

⇒ the p.d. across the variable resistor V_1 decreases.

When V_1 is less than V_2 the output of the op-amp V_o is positive.

When V_o is more than around 0·7 V the transistor is switched on.

The transistor then switches on the relay which switches on the heating circuit.

(b) The variable resistor is used to alter the temperature at which the heating circuit switches on.

Changing the resistance of the variable resistor changes the resistance of the thermistor at which V_o becomes big enough to switch on the transistor.

5.3 Radiation

Exercises

Exercise 33 Waves

1 (a) Frequency of the waves = 2·0 Hz.

(b) $f = 2{\cdot}0$ Hz $\qquad T = \frac{1}{f} = \frac{1}{2{\cdot}0} = 0{\cdot}50$ s

(c) Two waves pass the boy each second.

(d) $\lambda = 30$ cm $= 0{\cdot}30$ m $\qquad v = f\lambda$

$f = 2{\cdot}0$ Hz $\qquad = 2{\cdot}0 \times 0{\cdot}30 = 0{\cdot}60$ m s^{-1}

(e) $f = 3{\cdot}0$ Hz $\qquad \Rightarrow \quad 0{\cdot}60 = 3{\cdot}0 \times \lambda$

$\Rightarrow \quad \lambda = 0{\cdot}20$ m

2 (a) Wavelengths: Any three of BC, FH, GI, JL, KM and DE (A and B).

(b) Points in phase: Any four of F and H, G and I, J and L, K and M, D and E.

(c) Points out of phase: Any four of D and B, B and E, E and C, F and K, K and H, H and M, J and G, G and L, L and I (A and E, D and C, F and M, J and I).

3

Behaviour	Frequency	Period	Wavelength	Speed
Reflection	No change	No change	No change	No change
Refraction	No change	No change	Changes	Changes
Diffraction	No change	No change	No change	No change

4 (a) Audible sound waves have a much longer wavelength than visible light waves. The sound waves diffract around the tree so that you can hear your name being called. The light waves do not diffract sufficiently at the edges of the tree for you to be able to see your friend.

(b) Light is reflected from the surface of the wood. Light reflected from wood above the surface of the pond travels straight to your eye. Light reflected from wood below the surface is refracted as it moves from the water into the air. This changes the direction of the light so that it appears to come from a different place.

(c) Light waves are reflected from the book. The white pages reflect more light than the black ink of the words.

(d) The colour of the dress depends on the mixture of wavelengths of visible light reflected from the material. The mixture of wavelengths of visible light in daylight is different from the mixture of wavelengths of visible light in the light inside the shop.

5 (a) $f = 102{\cdot}5$ MHz $= 102{\cdot}5 \times 10^6$ Hz $\qquad T = \frac{1}{f} = \frac{1}{102{\cdot}5 \times 10^6}$

$= 9{\cdot}7561 \times 10^{-9} = 9{\cdot}756 \times 10^{-9}$ s

***Exercises** continued* ➢

Exercises continued

(b) $f = 102{\cdot}5 \times 10^6$ Hz $\qquad v = f\lambda$

$v = c = 3{\cdot}00 \times 10^8$ m s^{-1} $\qquad 3{\cdot}00 \times 10^8 = 102{\cdot}5 \times 10^6 \times \lambda$

$\lambda = ?$ $\qquad \lambda = 2{\cdot}927 = 2{\cdot}93$ m

(c) Frequency is not changed by reflection.

After reflection $f = 102{\cdot}5$ MHz.

Exercises

Exercise 34 Interference

1 (a) Constructive interference occurs at all points on the broken line. All of the points on this line are the same distance from P and Q ⇒ path difference is always zero.

(b) First find the path difference.

path difference = $QT - PT$

$PT = 0{\cdot}640$ m $\qquad = 0{\cdot}760 - 0{\cdot}640$

$QT = 0{\cdot}760$ m $\qquad = 0{\cdot}12$ m

Now find the number of wavelengths, n.

$\lambda = 4{\cdot}0$ cm $= 0{\cdot}040$ m $\qquad 0{\cdot}12 = n \times 0{\cdot}040$

$\Rightarrow \qquad n = 3$

⇒ constructive interference occurs at T

(c) There are three points of destructive interference between T and the broken line.

Including T and the broken line there are four points of constructive interference – for path differences = 3λ, 2λ, λ and 0 (i.e. 0λ). There is one point of destructive interference between each pair of adjacent points of constructive interference.

2 (a) Bright fringes: path difference = 0 nm, 600 nm, 1200 nm, 1800 nm, 2400 nm …

(b) Dark fringes: path difference = 300 nm, 900 nm, 1500 nm, 2100 nm, 2700 nm …

Exercises

Exercise 35 Gratings and spectra

1 (a) Grating has $2{\cdot}80 \times 10^5$ lines per metre $\Rightarrow d = \dfrac{1}{2{\cdot}80 \times 10^5} = 3{\cdot}571 \times 10^{-6}$ m.

$\theta = 25°$ $\qquad d \sin\theta = n\lambda$

$n = 3$ $\qquad 3{\cdot}571 \times 10^{-6} \times \sin 25° = 3 \times \lambda$

$\lambda = ?$ $\qquad \Rightarrow \qquad \lambda = 5{\cdot}031 \times 10^{-7} = 5{\cdot}03 \times 10^{-7}$ m

(b) $\qquad v = f\lambda$

$v = c = 3{\cdot}00 \times 10^8 \text{ m s}^{-1}$ $\qquad 3{\cdot}00 \times 10^8 = f \times 5{\cdot}03 \times 10^{-7}$

$\Rightarrow \qquad f = 5{\cdot}964 \times 10^{14} = 5{\cdot}96 \times 10^{14}$ Hz

(c) The colour of this light is green (the wavelength is approximately 500 nm).

2 First find the wavelength of the light.

$v = f\lambda$

$v = c = 3{\cdot}00 \times 10^8 \text{ m s}^{-1}$ $\qquad \Rightarrow \qquad 3{\cdot}00 \times 10^8 = 5{\cdot}20 \times 10^{14} \times \lambda$

$f = 5{\cdot}20 \times 10^{14}$ Hz $\qquad \Rightarrow \qquad \lambda = 5{\cdot}769 \times 10^{-7}$ m $\qquad$ *Intermediate value – four figures!*

$d = 4{\cdot}50 \times 10^{-6}$ m $\qquad d \sin\theta = n\lambda$

$n = 2$ $\qquad \Rightarrow \quad 4{\cdot}50 \times 10^{-6} \times \sin\theta = 2 \times 5{\cdot}769 \times 10^{-7}$

$\Rightarrow \qquad \sin\theta = 0{\cdot}2564$

$\Rightarrow \qquad \theta = 14{\cdot}86°$

The angle between the second order maxima = $2\theta = 29{\cdot}72° = 29{\cdot}7°$.

Both second-order maxima are deviated by angle θ from the zero order maximum.

3 The grating has $3{\cdot}2 \times 10^5$ lines per metre $\Rightarrow d = \dfrac{1}{3{\cdot}20 \times 10^5} = 3{\cdot}125 \times 10^{-6}$ m.

Assume λ at violet end = 400 nm = $4{\cdot}0 \times 10^{-7}$ m.

$d \sin\theta = n\lambda$

$n = 1$ $\qquad \Rightarrow \qquad 3{\cdot}125 \times 10^{-6} \times \sin\theta = 1 \times 4{\cdot}0 \times 10^{-7}$

$\Rightarrow \qquad \theta_{violet} = 7{\cdot}354°$

Assume λ at red end = 700 nm = $7{\cdot}0 \times 10^{-7}$ m

$\Rightarrow \qquad 3{\cdot}125 \times 10^{-6} \times \sin\theta = 1 \times 7{\cdot}0 \times 10^{-7}$

$\Rightarrow \qquad \theta_{red} = 12{\cdot}94°$.

The difference in the deviation = $12{\cdot}94° - 7{\cdot}354° = 5{\cdot}586 = 5{\cdot}59°$.

Exercises

Exercise 36 Refraction of light

1 (a) $v_{diamond} = 1{\cdot}24 \times 10^8$ m s^{-1} $\quad n = \frac{c}{v}$

$c = 3{\cdot}00 \times 10^8$ m s^{-1} $\Rightarrow$ $n = \frac{3{\cdot}00 \times 10^8}{1{\cdot}24 \times 10^8} = 2{\cdot}419 = 2{\cdot}42$

(b) $f = 4{\cdot}58 \times 10^{14}$ Hz $\quad v = f\lambda$

$\Rightarrow \quad 1{\cdot}24 \times 10^8 = 4{\cdot}58 \times 10^{14} \times \lambda$

$\Rightarrow \quad \lambda = 2{\cdot}707 \times 10^{-7} = 2{\cdot}71 \times 10^{-7}$ m

(c) The frequency in air = $4{\cdot}58 \times 10^{14}$ Hz.

2 (a) The light travels faster in air.

$\theta > 35° \Rightarrow \sin\theta > \sin 35° \Rightarrow n = \frac{\sin\theta}{\sin 35°} > 1$. Also $n = \frac{v_1}{v_2}$, $n > 1 \Rightarrow v_1 > v_2$.

(b) $\theta_2 = 35°$ $\quad n = \frac{\sin\theta_1}{\sin\theta_2}$

$n_{water} = 1{\cdot}33$ $\Rightarrow$ $1{\cdot}33 = \frac{\sin\theta_1}{\sin 35°}$ *value of n from data sheet!*

$\sin\theta_1 = 0{\cdot}763$

$\theta_1 = 49{\cdot}7° = 50°$

3 (a) Did you notice the incident ray is inside the plastic?

angle of incidence = $(90 - 58)° = 32°$ *i.e. angle between the ray and the normal!*

$n = 1{\cdot}38$ $\quad n = \frac{\sin\theta_1}{\sin\theta_2}$

$\Rightarrow \quad 1{\cdot}38 = \frac{\sin\theta_1}{\sin 32°}$

$\Rightarrow \quad \theta_1 = 46{\cdot}99 = 47°$

(b) $v = c = 3{\cdot}00 \times 10^8$ m s^{-1} $\quad v = f\lambda$ *First calculate wavelength in a vacuum.*

$f = 4{\cdot}7 \times 10^{14}$ Hz $\quad 3{\cdot}00 \times 10^8 = 4{\cdot}70 \times 10^{14} \times \lambda$

$\lambda = 6{\cdot}383 \times 10^{-7}$ m *intermediate value – extra figure!*

$n = 1{\cdot}38$ $\quad n = \frac{\lambda_1}{\lambda_2}$ *Now calculate wavelength in plastic.*

Exercises continued ➢

Exercises continued

$\lambda_1 = 6{\cdot}383 \times 10^{-7}$ m $\qquad 1{\cdot}38 = \dfrac{6.383 \times 10^{-7}}{\lambda_2}$

$$\lambda_2 = 4{\cdot}625 \times 10^{-7} = 4{\cdot}63 \times 10^{-7}\ \text{m}$$

(c) The light is red (λ in air > *620* nm).

(d) $$n = \frac{c}{v}$$

$$\Rightarrow \quad 1{\cdot}38 = \frac{3.00 \times 10^8}{v_2}$$

$$\Rightarrow \quad v_2 = 2{\cdot}174 \times 10^8 = 2{\cdot}17 \times 10^8\ \text{m s}^{-1}$$

4 Red light: $\theta_1 = (90 - 60)° = 30°$ $\qquad n = \dfrac{\sin \theta_1}{\sin \theta_2}$

$n = 1{\cdot}48$ $\qquad \Rightarrow \quad 1{\cdot}48 = \dfrac{\sin 30}{\sin \theta_2}$

$$\Rightarrow \quad \theta_2 = 19{\cdot}75°$$

Violet light: $\theta_1 = 30°$ $\qquad n = \dfrac{\sin \theta_1}{\sin \theta_2}$

$n = 1{\cdot}52$ $\qquad \Rightarrow \quad 1{\cdot}52 = \dfrac{\sin 30}{\sin \theta_2}$

$$\Rightarrow \quad \theta_2 = 19{\cdot}20°$$

Angle between red and violet ends of the spectrum = $(19{\cdot}75 - 19{\cdot}20)° = 0{\cdot}55°$.

Exercises

Exercise 37 Total internal reflection and critical angle

1 $n = 1{\cdot}31$ $\qquad n = \dfrac{1}{\sin \theta_c}$

$$\Rightarrow \quad 1{\cdot}31 = \frac{1}{\sin \theta_c}$$

$$\Rightarrow \quad \sin \theta_c = \frac{1}{1.31}\ (= 0{\cdot}7634)$$

$$\theta_c = 49{\cdot}76 = 49{\cdot}8°$$

Exercises continued ➢

Exercises continued

2 (a) $n_{glass} = \dfrac{c}{v_{glass}} \Rightarrow 1{\cdot}46 \times v_{glass} = c$ **and** $n_{water} = \dfrac{c}{v_{water}} \Rightarrow 1{\cdot}30 \times v_{water} = c$

$$\Rightarrow \quad 1{\cdot}30 \times v_{water} = 1{\cdot}46 \times v_{glass}$$

$$\Rightarrow \quad n = \frac{v_{water}}{v_{glass}} = \frac{1{\cdot}46}{1{\cdot}30}$$

$$= 1{\cdot}123 = 1{\cdot}12$$

(b)

$$\sin\theta_c = \frac{1}{n} = \frac{1}{1{\cdot}12}$$

$$\Rightarrow \quad \theta_c = 63{\cdot}23 = 63{\cdot}2°$$

Exercises

Exercise 38 Irradiance

1 $I_1 = 3{\cdot}6\ \text{W m}^{-2}$ $\quad Id^2 = k \quad \Rightarrow 3{\cdot}6 \times 2{\cdot}0^2 = I_2 \times 6{\cdot}0^2$ *Careful with the substitution!*

$d_1 = 2{\cdot}0\ \text{m} \quad \Rightarrow \quad I_2 = 0{\cdot}40\ \text{W m}^{-2}$

$d_2 = 6{\cdot}0\ \text{m}$

2 The irradiance due to the refracted ray may vary with distance from point X.

Exercises

Exercise 39 Photoelectric effect

1 (a) Minimum photon energy to release an electron = work function = $9{\cdot}00 \times 10^{-19}$ J.

(b) $f = 1{\cdot}51 \times 10^{15}$ Hz

$$E = hf$$

$$= 6{\cdot}63 \times 10^{-34} \times 1{\cdot}51 \times 10^{15}$$

$$= 1{\cdot}001 \times 10^{-18} = 1{\cdot}00 \times 10^{-18}\ \text{J}$$

(c) maximum kinetic energy of photo electron $= hf - hf_o$

$$= 1{\cdot}00 \times 10^{-18} - 9{\cdot}00 \times 10^{-19}$$

$$= 1{\cdot}00 \times 10^{-19}\ \text{J}$$

(d) $m_e = 9{\cdot}11 \times 10^{-31}$ kg $\quad E_k = ½mv^2$

$$\Rightarrow \quad 1{\cdot}00 \times 10^{-19} = ½ \times 9{\cdot}11 \times 10^{-31} \times v^2$$

$$\Rightarrow \quad v^2 = 2{\cdot}195 \times 10^{11}$$

$$\Rightarrow \quad v = 4{\cdot}685 \times 10^5 = 4{\cdot}69 \times 10^5\ \text{m s}^{-1}$$

Exercises continued ➢

Exercises continued

2 (a) At double the distance, irradiance = ¼ × initial irradiance

$\Rightarrow$ number of photons per second incident on the surface = ¼ × initial number

$\Rightarrow$ number of photoelectrons ejected per second = ¼ × initial number ejected per second.

(b) Neither the photon energy nor the work function of the surface changes

$\Rightarrow$ no change to the maximum kinetic energy of the photoelectrons

$\Rightarrow$ no change to the maximum velocity of the photoelectrons.

3 (a) $f = 1{\cdot}25 \times 10^{15}$ Hz

$E = hf$

$= 6{\cdot}63 \times 10^{-34} \times 1{\cdot}25 \times 10^{15}$

$= 8{\cdot}288 \times 10^{-19}$ J *Note this is not a final answer!*

No photoelectrons are ejected as the photon energy is less than the work function.

(b) The number of photoelectrons ejected per second remains zero. Photon energy is still less than the work function.

(c) $f = 1{\cdot}85 \times 10^{15}$ Hz

$E_k = hf - hf_o$

$= (6{\cdot}63 \times 10^{-34} \times 1{\cdot}85 \times 10^{15}) - 9{\cdot}20 \times 10^{-19}$

$= 3{\cdot}066 \times 10^{-19} = 3{\cdot}07 \times 10^{-19}$ J

Exercises

Exercise 40 Energy levels in atoms

1 (a) 5 (E_5 to E_0, E_4 to E_0, E_3 to E_0, E_2 to E_0 and E_1 to E_0).

(b) 15 (5 to E_0 + 4 to E_1 + 3 to E_2 + 2 to E_3 + 1 to E_4).

(c) E_5 to E_0 produces photons with the shortest wavelength.

shortest wavelength $\Rightarrow$ highest frequency (using $v = f\lambda$)

$\Rightarrow$ biggest energy ($E = hf$) – E_5 to E_0 is the biggest energy change

2 (a) Level A.

(b) $W_2 = -4{\cdot}00 \times 10^{-20}$ J

$W_1 = -2{\cdot}21 \times 10^{-18}$ J

$E = W_2 - W_1$

$= [-4{\cdot}00 \times 10^{-20} - (-2{\cdot}21 \times 10^{-18})]$

$= 2{\cdot}17 \times 10^{-18}$ J

(c) $h = 6{\cdot}63 \times 10^{-34}$ J s

$E = hf$

$\Rightarrow \quad 2{\cdot}17 \times 10^{-18} = 6{\cdot}63 \times 10^{-34} \times f$

$\Rightarrow \quad f = 3{\cdot}273 \times 10^{15} = 3{\cdot}27 \times 10^{15}$ Hz

Exercises continued ➢

Exercises continued

(d) $W_2 = -1.80 \times 10^{-19}$ J $\quad hf = W_2 - W_1$

$W_1 = -6.23 \times 10^{-19}$ J $\Rightarrow \quad 6.63 \times 10^{-34} \times f = [-1.80 \times 10^{-19} - (-6.23 \times 10^{-19})]$

$\Rightarrow \quad f = 6.682 \times 10^{14}$ Hz

$c = 3.00 \times 10^8$ m s^{-1} $\quad v = f\lambda$

$\Rightarrow \quad 3.00 \times 10^8 = 6.682 \times 10^{14} \times \lambda$

$\Rightarrow \quad \lambda = 4.489 \times 10^{-7} = 4.49 \times 10^{-7}$ m

(e) The photon is visible.

The wavelength is in the range of the visible spectrum, 400 nm $< \lambda <$ 700 nm.

(f) The student sees the complete spectrum.

The energies for the dark lines in its absorption spectrum coincide exactly with the energies for the bright lines in its emission spectrum. When the two spectra overlap on the same scale the bright lines of the emission spectrum cover the dark lines of the absorption spectrum.

Exercises

Exercise 41 Lasers

1 Spontaneous emission is random. An electron in an excited state falls to a lower energy level at a time that cannot be predicted. The photon emitted may have a number of possible energies.

In stimulated emission an incident photon causes an electron in an excited state to fall to a specific lower energy level and emit another photon. The emitted photon is in phase with the incident photon. The energy of the emitted photon is equal to the energy of the incident photon.

2 The mirrors in a laser reflect photons back into the material of the laser. This enables further stimulated emissions to occur and ensures that the light beam gains more energy than it loses.

3 (a) $P = 0.825$ W $\quad I = \dfrac{P}{A}$

$A = 1.21$ mm^2

$= 1.21 \times 10^{-6}$ m^2 $\quad = \dfrac{0.825}{1.21 \times 10^{-6}}$

$= 6.818 \times 10^5 = 6.82 \times 10^5$ W m^{-2}

(b) Trebling the distance between the laser and the screen has little or no effect on the irradiance due to the laser beam.

Exercises

Exercise 42 Semiconductors

1 (a) n-type silicon is formed.

(b) The semiconductor material is electrically neutral.

The phosphorus atoms added to the crystal are electrically neutral – they have an equal number of protons and electrons.

2 The majority charge carriers are holes. (The material formed is p-type silicon.)

Exercises

Exercise 43 Semiconductor devices

1 (a)

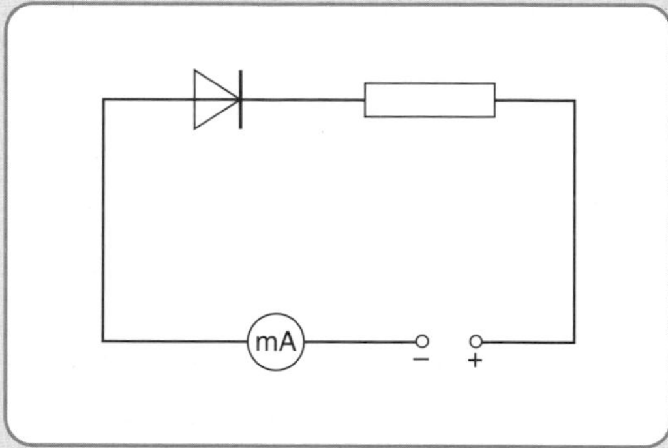

(b) First calculate the change in irradiance of the radiation incident on the photodiode.

$Id^2 = k \quad \Rightarrow \quad I_1 \times 1{\cdot}50^2 = I_2 \times 2{\cdot}50^2$

$\Rightarrow \quad I_2 = 0{\cdot}360 \times I_1$

Leakage current is directly proportional to irradiance

$\Rightarrow$ new reading on meter $= 0{\cdot}360 \times 0{\cdot}280$

$= 0{\cdot}100{\cdot}8 = 0{\cdot}101$ mA

(c) When the lamp is further from the photodiode fewer photons per second are incident on the p-n junction

$\Rightarrow$ fewer positive and negative charges are produced each second

$\Rightarrow$ current is less.

2 (a) $\lambda = 595$ nm $= 5{\cdot}95 \times 10^{-7}$ m $\qquad v = f\lambda$

$c = 3{\cdot}00 \times 10^8$ m s^{-1} $\Rightarrow \quad 3{\cdot}00 \times 10^8 = f \times 5{\cdot}95 \times 10^{-7}$

$\Rightarrow \quad f = 5{\cdot}042 \times 10^{14}$ Hz

$E = hf$

Exercises *continued* ➢

Exercises continued

$h = 6{\cdot}63 \times 10^{-34}$ J s $\Rightarrow$ $E = 6{\cdot}63 \times 10^{-34} \times 5{\cdot}042 \times 10^{14}$

$= 3{\cdot}343 \times 10^{-19} = 3{\cdot}34 \times 10^{-19}$ J

(b) $Q = 1{\cdot}60 \times 10^{-19}$ C $\quad E = QV$

$\Rightarrow 3{\cdot}34 \times 10^{-19} = 1{\cdot}60 \times 10^{-19} \times V$

$\Rightarrow V = 2{\cdot}0875 = 2{\cdot}09$ V

(c) At the p-n junction, holes recombine with electrons. The electrons fall to lower energy levels and photons are emitted.

3 (a) When A is positive, C is negative. The diodes in branches AD (positive connected to p-type) and CB (negative connected to n-type) are forward-biased and conduct. The diodes in branches AB and CD are reverse-biased and do not conduct.

Electrons flow C → B → D → A.

When A is negative, C is positive. The diodes in branches AB and CD are forward-biased and conduct. The diodes in branches AD and BC are reverse-biased and do not conduct.

Electrons flow A → B → D → C.

In both cases the electron flow through the resistor is from B to D.

(This is called full-wave rectification – all of the a.c. energy is converted to d.c.)

(b) Connect a capacitor in parallel with the resistor to smooth the p.d. across the resistor.

4 Initially the readings on both meters are zero. The MOSFET is off.

As the resistance of the variable resistor is gradually increased the reading on the voltmeter increases. The reading on the milliammeter remains zero.

When the variable resistor has value 4 kΩ the reading on the voltmeter is 2 V. The MOSFET switches on at around this point.

As the resistance of the variable resistor is increased from 4 kΩ to 14 kΩ the reading on the milliammeter gradually increases from zero. The voltmeter reading increases from 2 V to 4·5 V.

Exercises

Exercise 44 Nuclear reactions

1 Rutherford's experiment was carried out in a vacuum so that only gold atoms would be in the path of the alpha particles. This ensures that the observed results are due entirely to interactions between alpha particles and gold atoms.

Exercises continued ➢

Exercises *continued*

2 (a) $^{238}_{92}U \rightarrow \ldots \rightarrow\, ^{222}_{86}Rn$. Total change in mass number = 238 – 222 = 16.

Mass number of 1 alpha particle = 4 ⇒ number of alpha particles emitted = $\frac{16}{4}$ = 4.

(b) Total change in atomic number due to alpha emission = 8.

Actual total change in atomic number = 92 – 86 = 6.

Change in atomic number due to beta emission = +1

⇒ number of beta particles emitted = 2.

(c) Neither mass number nor atomic number change when a gamma photon is emitted. It is not possible to calculate the number of gamma photons emitted from the data given.

3 (a) Z = 57 ⇒ there are 57 protons in the La nucleus.

(b) A = 95 number of neutrons in Nb nucleus = $A - Z$

Z = 41 = 95 – 41 = 54

(c) Mass number is conserved: total mass number of reactants = 235 + 1 = 236.

Total mass number of products listed = 139 + 95 = 234.

Difference = 236 – 234 = 2 ⇒ number of neutrons released = 2.

(**Note:** Mass number of electrons = 0.)

(d) Atomic number is conserved: total atomic number of reactants = 92 + 0 = 92.

Total atomic number of products listed = 57 + 41 = 98.

Difference = 92 – 98 = – 6 ⇒ number of electrons released = 6.

(**Note:** Atomic number of neutrons = 0.)

4 (a) This is an induced fission reaction.

The uranium nucleus absorbs a neutron before splitting.

(b) Total mass of reactants $= 3{\cdot}902\,15 \times 10^{-25} + 1{\cdot}674\,93 \times 10^{-27}$

$= 3{\cdot}918\,90 \times 10^{-25}$ kg.

Total mass of products $= 2{\cdot}389\,21 \times 10^{-25} + 1{\cdot}476\,07 \times 10^{-25} + (3 \times 1{\cdot}674\,93 \times 10^{-27})$

$= 3{\cdot}915\,53 \times 10^{-25}$ kg

⇒ mass lost $= 3{\cdot}37 \times 10^{-28}$ kg

$E = mc^2$

⇒ $E = 3{\cdot}37 \times 10^{-28} \times (3{\cdot}00 \times 10^8)^2$

$= 3{\cdot}033 \times 10^{-11} = 3{\cdot}03 \times 10^{-11}$ J

(c) When the neutrons are released they are moving too fast to cause further fission reactions. The neutrons are slowed down by being passed them through a moderator. When they have slowed sufficiently, further uranium-235 nuclei may absorb neutrons leading to further fissions.

Exercises *continued* ➢

Exercises continued

5 (a) Total mass of reactants $= 3{\cdot}34449 \times 10^{-27} + 5{\cdot}00826 \times 10^{-27}$

$= 8{\cdot}35275 \times 10^{-27}$ kg.

Total mass of products $= 6{\cdot}64647 \times 10^{-27} + 1{\cdot}67493 \times 10^{-27}$

$= 8{\cdot}32140 \times 10^{-27}$ kg

$\Rightarrow$ mass lost $= 3{\cdot}135 \times 10^{-29}$ kg

$E = mc^2$

$\Rightarrow E = 3{\cdot}135 \times 10^{-29} \times (3{\cdot}00 \times 10^8)^2$

$= 2{\cdot}8215 \times 10^{-12} = 2{\cdot}82 \times 10^{-12}$ J.

(b) Energy required per second $= 4{\cdot}5 \times 10^6$ J

$\Rightarrow$ number of reactions required each second $= \dfrac{4{\cdot}5 \times 10^6}{2{\cdot}82 \times 10^{-12}}$

$= 1{\cdot}595 \times 10^{18} = 1{\cdot}6 \times 10^{18}$ reaction/s.

Exercises

Exercise 45 Dosimetry

1 $N = 12\,000$ $\quad A = \dfrac{N}{t}$

$t = 5$ minutes $= 300$ s $\Rightarrow \quad = \dfrac{12\,000}{300}$

$= 40$ Bq

2 Mass of mineral in the rock $= 0{\cdot}15 \times 2{\cdot}4 = 0{\cdot}36$ kg.

Activity of 100 g (i.e. 0·100 kg) $= 2{\cdot}3 \times 10^4$ Bq

$\Rightarrow$ Activity of rock $= 2{\cdot}3 \times 10^4 \times \dfrac{0{\cdot}36}{0{\cdot}100}$

$= 8{\cdot}28 \times 10^4 = 8{\cdot}3 \times 10^4$ Bq.

3 (a) $H = 3{\cdot}75 \times 10^{-2}$ Sv $\quad \dot{H} = \dfrac{H}{t}$

$t = 125 \times 12 = 1500$ h $\Rightarrow \quad = \dfrac{3.75 \times 10^{-2}}{1500}$

$= 2{\cdot}50 \times 10^{-5}$ Sv h^{-1}

Exercises continued ➢

Exercises continued

(b) $w_R = 4$

$$\dot{H} = \dot{D}w_R$$
$$\Rightarrow 3{\cdot}75 \times 10^{-2} = \dot{D} \times 4$$
$$\Rightarrow \quad \dot{D} = 9{\cdot}375 \times 10^{-3} \text{ Gy/year}$$
$$= 7{\cdot}813 \times 10^{-4} = 7{\cdot}81 \times 10^{-4} \text{ Gy/month}$$

4 (a) Gamma rays:

$m = 150 \text{ g} = 0{\cdot}150 \text{ kg}$ $\qquad H = Dw_R = \frac{E}{m}w_R$

$E = 46 \text{ mJ} = 0{\cdot}046 \text{ J}$ $\qquad = \frac{0{\cdot}046}{0{\cdot}150} \times 1$

$w_R = 1$ $\qquad = 0{\cdot}307 \text{ Sv}$

Fast neutrons:

$m = 320 \text{ g} = 0{\cdot}320 \text{ kg}$ $\qquad H = \frac{E}{m}w_R$

$E = 12 \text{ mJ} = 0{\cdot}012 \text{ J}$ $\qquad = \frac{0{\cdot}012}{0{\cdot}320} \times 8$

$w_R = 8$ $\qquad = 0{\cdot}30 \text{ Sv}$

Alpha particles:

$m = 260 \text{ g} = 0{\cdot}260 \text{ kg}$ $\qquad H = \frac{E}{m}w_R$

$E = 6{\cdot}5 \text{ mJ} = 0{\cdot}0065 \text{ J}$ $\qquad = \frac{0{\cdot}0065}{0{\cdot}260} \times 20$

$w_R = 20$ $\qquad = 0{\cdot}50 \text{ Sv}$

The alpha particles carry the greatest risk of biological harm.

(b) The risk to health also depends on the susceptibilities to harm of the irradiated tissues.

Exercises

Exercise 46 Radiation and safety

1 (a) $D = 4{\cdot}8 \times 10^{-5} \text{ Gy}$ $\qquad D = \frac{E}{m}$

$m = 0{\cdot}65 \text{ kg}$ $\qquad \Rightarrow 4{\cdot}8 \times 10^{-5} = \frac{E}{0{\cdot}65}$

Exercises continued ➢

Exercises continued

$\Rightarrow \quad E = 3{\cdot}12 \times 10^{-5} = 3{\cdot}1 \times 10^{-5}$ J

(b) $D_1 = 4{\cdot}8 \times 10^{-5}$ Gy $\quad \Rightarrow \quad \dfrac{D_1}{D_2} = \dfrac{4{\cdot}8 \times 10^{-5}}{3{\cdot}0 \times 10^{-6}} = 16$

$D_2 = 3{\cdot}0 \times 10^{-6}$ Gy $\quad \Rightarrow$ the lead has reduced the absorbed dose by a factor of 16

$16 = 2 \times 2 \times 2 \times 2 = 2^4 \quad \Rightarrow$ 120 mm is equivalent to four half-value thicknesses of lead

$$\text{half-value thickness} = \frac{120}{4} = 30 \text{ mm}$$

(**Note:** This is an exception – it is easier not to change to m for this calculation.)

2 Radioactive source is a point source of radiation $\Rightarrow$ use inverse square law $Id^2 = k$.

$d_1 = 4{\cdot}5$ m $\quad \Rightarrow \quad 50 \times 4{\cdot}5^2 = \text{count rate} \times 1{\cdot}5^2$

$d_2 = 1{\cdot}5$ m $\quad \Rightarrow$ count rate = 450 counts per minute

5.4 Course skills

Exercises

Exercise 47 Units, prefixes and scientific notation

1 (a) 40 000 V = $4{\cdot}0 \times 10^4$ V

(b) 0·000 036 5 W m^{-2} = $3{\cdot}65 \times 10^{-5}$ W m^{-2}

(c) 200 pF = 200×10^{-12} F = $2{\cdot}00 \times 10^{-10}$ F

(d) 4500 km = 4500×10^3 m = $4{\cdot}500 \times 10^6$ m

(e) 5·0 GHz = $5{\cdot}0 \times 10^9$ Hz

(f) 1·25 μA = $1{\cdot}25 \times 10^{-6}$ A

2 (a) $9{\cdot}0 \times 10^{-2}$ kg m^{-3} = 0·090 kg m^{-3}

(b) 724 mm = 0·724 m

(c) $0{\cdot}098 \times 10^2$ m s^{-2} = 9·8 m s^{-2}

(d) 8·8 MJ = 8 800 000 J

(e) $3{\cdot}16 \times 10^{-5}$ Sv = 0·000 031 6 Sv

Exercises

Exercise 48 Uncertainties

1 (a) $I = 6{\cdot}37$ A $\qquad E_h = ItV = cm\Delta T$

$V = 9{\cdot}41$ V $\quad \Rightarrow 6{\cdot}37 \times 1200 \times 9{\cdot}41 = c \times 1{\cdot}15 \times 25$

Exercises continued ➢

Exercises continued

t = 1200 s ⇒ c = 2502

m = 1·15 kg = 2500 J kg^{-1} °C^{-1}

ΔT = 25 °C

(b) Work out the percentage uncertainty in each measurement.

mass of liquid: percentage uncertainty = $\frac{0.01}{1.15} \times 100\%$ = 0·87% = 0·9%

heater voltage: percentage uncertainty = $\frac{0.02}{9.41} \times 100\%$ = 0·21% = 0·2%

heater current: percentage uncertainty = $\frac{0.02}{6.37} \times 100\%$ = 0·31% = 0·3%

time: percentage uncertainty = $\frac{2}{1200} \times 100\%$ = 0·17% = 0·2%

temperature rise: percentage uncertainty = $\frac{1}{25} \times 100\%$ = 4%

⇒ approximate percentage uncertainty in final numerical value = 4%

⇒ absolute uncertainty = $\frac{4}{100} \times 2500$ = 100 J kg^{-1}°C^{-1}

Student's result is c = 2500 J kg^{-1}°C^{-1} ± 100 J kg^{-1}°C^{-1}.

2 (a)

Angle in air θ_1/°	20·0	30·0	40·0	50·0	60·0	70·0
Angle in plastic θ_2/°	13·5	19·5	25·5	31·0	36·0	39·5
$\frac{\sin\theta_1}{\sin\theta_2} = n$	1·465	1·498	1·493	1·487	1·473	1·477

mean value of n = (1·465 + 1·498 + 1·493 + 1·487 + 1·473 + 1·477)/6

= 1·482

(b) Uncertainty in mean = $\frac{\text{maximum value} - \text{minimum value}}{n}$

= $\frac{1.498 - 1.465}{6}$

= 0·005

(c) The pair of readings θ_1 = 70°, θ_2 = 39·5° is likely to give the most accurate value for n.

The absolute uncertainty is the same for each angle measured

⇒ the percentage uncertainties are smallest for the largest angles

⇒ percentage uncertainty is smallest in the value of n calculated for the largest angles.